WRITER'S GUIDE

Life Sciences

ARTHUR W. BIDDLE
University of Vermont

DANIEL J. BEAN
St. Michael's College

with TOBY FULWILER
University of Vermont

D. C. HEATH AND COMPANY
Lexington, Massachusetts *Toronto*

Published simultaneously in Canada.

Printed in the United States of America.

International Standard Book Number: 0-669-12003-0

Library of Congress Catalog Card Number: 86-81358

Preface

The best way to learn biology is by writing it — that's the principle behind this book. *Writer's Guide: Life Sciences* applies current writing theory to the special needs of this exciting discipline. The result is a powerful aid for students of the life sciences at every level.

TO THE STUDENT

Whether you are a committed biology major or a chemist or English major taking an elective, whether you are enrolled in your first life science course or your last, if you want to understand this challenging field, *Writer's Guide* is meant for you. This book shows you how to master library research tools, explains why the research proposal is so important (and teaches you how to write one), even demystifies the Abstract. You'll find help with all your writing needs in the life sciences.

TO THE INSTRUCTOR

Written by a biologist and a writing specialist, *Writer's Guide: Life Sciences* offers a variety of resources adaptable to virtually any course in your curriculum. Topics and writing assignments are sequenced from the less to the more sophisticated. Thus introductory students might find the first three or four chapters most valuable, leading up to an article summary, a book review, or, more ambitiously, a term paper based on library research. Those same chapters would also prove useful for intermediate level students, but their major piece of writing might be a research proposal and perhaps a research report, as explained in Chapters 5 and 6. Advanced students, too, would probably write a research proposal and report. Every student required to do library work will find the chapter on library research and materials goes beyond a mere introduction to the building. We show search strategies and explain how to use the difficult, but important, reference sources. Keeping a journal (as explained in Chapter 2) seems to bring out the best in just about every student from freshman to graduate. Finally, concise guides to usage and punctuation provide handy reference aids.

Each chapter of *Writer's Guide* is designed to be self-instructional. Although the value of many assignments would be enhanced by class discussion or individual conference, students can use this book independently. In the chapter on the term paper, for instance, the reader learns how to select a topic, then how to analyze and research that topic, and finally how to organize and present the material. Each step culminates in a writing assignment, each assignment leads to the next, until the student has produced a finished essay. Undoubtedly, some instructors will simply assign a chapter as a means of assigning a paper. However you choose to use this book, you will find that it improves your students' understanding of the course material, not just their writing ability.

Acknowledgements

The authors express their appreciation to all those who assisted in the preparation of this book. Our greatest debt is owed to our students. In English classes as well as in life science courses, students field-tested most of this material and gave very practical advice for improvements. The majority of writing samples in this book came out of those classes. We're grateful to the authors for permission to use their work.

We also appreciate the help of friends and colleagues. At the University of Vermont Lynne Bond, Virginia Clark, Mary Jane Dickerson, Toby Fulwiler, Littleton Long, Anthony Magistrale, and Henry Steffens have been especially supportive. Biologists and writing specialists at other colleges and universities aided too: Joan Edelstein, San Jose State University; Ronald Stecker, San Jose State University; and Ruth Wilson, California State University — San Bernardino. For this assistance, we give thanks.

Arthur W. Biddle
Daniel J. Bean
Burlington, Vermont

Contents

[1] *Writing in the Life Sciences*

PREVIEW: *Writing is a powerful way of learning. This chapter explains the value of writing in biology courses and tells the decisions you need to make before starting out.*

WHY WRITE?

That's a pretty good question. You enrolled in a biology course, not a writing workshop, but now you find that you're expected to produce a lot of writing. Why? Here are some very good reasons.

Writing will help you learn science. You've probably discovered the principle behind this fact already: we learn best, not as passive recipients of lectures and textbooks, but as active participants, making meaning for ourselves. Writing is one of the best ways to get involved in your own education. That's what this book is all about — writing to learn. Your personal involvement through writing will lead you to fuller understanding of the life sciences.

Writing clarifies your understanding of the subject. Let's say you read a chapter in your textbook or listen to a fifty-minute lecture on cell structure and understand most of it. Writing what you comprehend helps you review, organize, and remember the material. But some of the information still puzzles you. By putting your questions on paper, by writing about your confusion, you begin to see just where the difficulty lies. Often, you can write your way to understanding. Even if that doesn't work, you'll know which sections of the chapter to reread or which notes to review. You can ask your professor an intelligent question: for instance, what are the biochemical differences between mitochondria and the cells they occupy?

Writing reveals your attitude toward a subject. An assignment might be to read two articles about acid precipitation and its effects on forests. As you study one of the essays, you agree with that writer's position. Then you read another piece and are persuaded to that point of view. Sound familiar? Professional writers can be persuasive — that's their line of work. That we readers sometimes have trouble assessing what we read shouldn't surprise us. What should we do about this? Listing the pros and cons of a given policy helps you see the strengths and weaknesses of each side. Writing can help you discover how you feel about the issue. Then you're on the way to defining your own position.

Writing helps you synthesize large amounts of information. The human mind is a marvel unduplicated by the most advanced computer. Still, most of us don't seem to command the kind of memory we need. Making notes supplements memory and provides access to limitless information. That's why note-taking is an essential part of research. Further, you can understand how the information you have collected is related by writing about it. Writing allows you to discover new ways of solving a problem.

Writing organizes your thoughts. You already know that. When you have a lot to accomplish, you make a list of "things to do today." When you prepare a speech or a class presentation, you jot down main ideas, then reorganize them into some sort of meaningful pattern. Combining invisible thoughts with the physical activity of forming words on paper helps you to see what you're thinking. And somehow the need to commit your thinking to the page focusses the energy of your mind. That sounds sort of mystical, but that's what seems to happen.

Learning through writing may seem to take longer, but you'll find that it leads to fuller understanding of the subject. And you'll like your new-found control of your studies. You've taken charge!

YOU AND THE WRITING PROCESS

Has this ever happened to you? Your professor assigns a paper, due at the end of the semester. You're not told much more about it — perhaps you get a list of acceptable topics or learn how many pages to write. Then, despite your best intentions, you wait until a couple of days before the due date to get started. This is an old and sad story.

There are better ways of doing things. Whether you need to write a term paper, a seminar presentation, or a book review — virtually any communication, in fact — the most effective means is the **process approach.** Using this method of composition, you work your way through three broad stages: **prewriting, drafting,** and **revising.** Most experienced writers work this way. And writers-in-training seem to make the greatest improvement when they practice composing in this fashion.

Prewriting

All the preparations the writer makes before starting to draft — that's what we mean by prewriting. Among these preparations are finding a topic, limiting that topic to manageable size, defining purpose, assessing audience, choosing a point of view, researching or interviewing, and taking notes. This prewriting stage of the process is much more crucial than many realize. When you know that, you're ahead of the rest. And when you master these preparations, you win a new control over your writing. You'll take the first steps in the next section, The Writer's Decisions, and in Chapter 2, Keeping a Journal. You will also find help with prewriting throughout this book as you learn how to jot down plans and outlines and to write a discovery draft.

Drafting

The second stage of the composing process, drafting, is what most people have in mind when they think of writing. Drafting is getting the words down on paper, much easier when you use the process approach. Chapters 3, 4, and 5 will guide you through this stage.

Revising

Revising, the third stage, involves much more than most writers-in-training suspect. Example: this chapter is now in its fifth draft. In other words, it has been revised four times. That's why professional writers have such big wastebaskets: they keep working on a piece until it's right. If you were to look up the word *revise* in the dictionary, you would find that it comes from the Latin *revidere* — to see again. True revision means just that, seeing again, looking once more at a draft with a willingness to consider changes, often big changes. You'll learn more about making these changes as you proceed through this book. Then, after you've revised your way to a good piece of work, refer to Chapter 9, A Concise Guide to Usage, and Chapter 10, Make Punctuation Work for You, for help with editing. Editing and proofreading are the final steps before submitting your work to reader or editor.

THE WRITER'S DECISIONS

During the prewriting phase, before beginning to draft, the writer confronts several questions: Why am I writing this? Who's going to read it? What will they be expecting? How should my voice sound? Consciously or not, writers must answer these questions each time they sit down to write. Whether you are researching a term paper for your biology course, or applying to graduate school, or writing a textbook for a course in writing in the life sciences, the questions are the same. Only the answers are different.

What is this piece about?

Your answer to this question establishes the **subject,** the true topic of this piece of writing.

Why am I writing this? What do I want this to do?

In answering these questions you make decisions about

purpose. **Purpose** is your intent, the reason that moves you to write and the desired result of that effort.

Who am I writing this for?

The answer to this question identifies your audience. **Audience** is the reader or readers you are addressing.

Who am I as I write this?

The answer to this question describes your voice. **Voice** is the character, personality, and attitudes you project toward your subject, toward your purpose, and toward your audience.

Subject, purpose, audience, and voice are controls in any job of writing. Once you make decisions or accept conditions concerning their natures, you establish certain parameters. Style, tone, readability, even organization and use of examples, are all governed by these initial choices.

The figure below is an attempt to show how the four decisions relate to each other. At the heart of the figure is subject, the focus of any piece of writing and usually the writer's first decision. The figure suggests the influence that subject has on purpose, audience, and voice, as well as the relationship they have to one another.

Decision 1: Subject

Finding something to write about is often the hardest of the pre-writing tasks for writers-in-training. In the "real world," of course, you would write only when you had something to say, perhaps to yourself in your journal or perhaps when you felt the need to express it to others. In college courses, however, you are often told to write, whether you need to or not. That's because your professors view the writing process as a means of learning. One objective in most biology courses is learning to perceive the world as a scientist. That means making the observations, asking the questions, and designing the experiments in the same ways that a practicing scientist does.

One of the best solutions to the problem of finding a **subject** is to anticipate it. Keeping a biology journal right from your first day

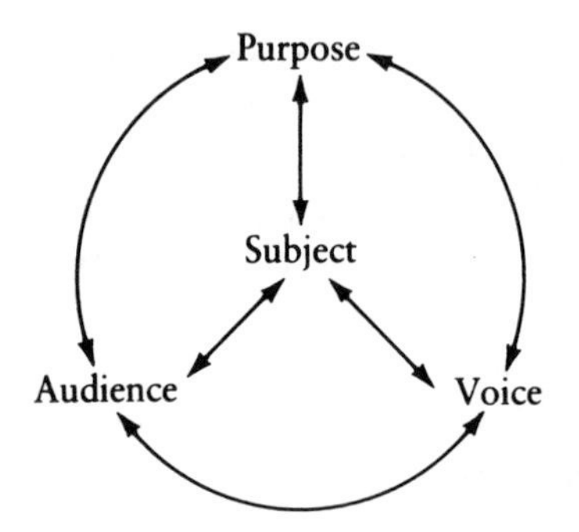

Figure 1.1 *The writer's decisions.*

in the general biology course will provide you with dozens of ideas for writing and research. In Chapter 2, Keeping a Journal, you'll learn how it is done.

Whatever subject you choose to write about should meet the following specifications:

1. It should fit the assignment. Does the subject fall within the scope of the course? A paper on the chemical pathways in C3 *vs.* C4 plants would be appropriate for a biochemistry course. A paper on vertebrate morphology would be inappropriate for the same course. Does the paper come within the limits established by the assignment? An account of Rachel Carson's *Silent Spring* would not fit the assignment, "Write a paper about a *current* issue of environmental concern." A study of the effects of acid rain on soil microorganisms would fit that assignment.

2. It should be of interest to you. This is an obvious criterion, but one that students often overlook. If you are to spend thirty or forty hours researching and writing a paper, you should feel some intellectual excitement about the subject. Of course, you might discover a new interest as you read about a topic you thought dull at first. But to the extent possible, begin with a question you genuinely want to answer.

3. It should be limited to allow adequate depth and breadth of coverage. If the assignment calls for a six- to eight-page paper, a topic like the structure and function of eukaryotic cells is doomed to failure. Trying to cover all aspects of both plant and animal cells and their variations is hopeless. Limiting that broad topic could yield some very workable subjects, though: one might be the structure of prokaryotic cells *vs.* eukaryotic cells. Another possibility is types and functions of muscle cells, and a third is special plant cell adaptations.

WRITING 1.1: FINDING A SUBJECT. Write a list of five topics suitable for a six- to eight-page paper from the broad subject: cell structure and function. Use the library if you need to.

Decision 2: Purpose

The **purpose** of a piece of writing can be complex, for it includes both the reason that moves you to write and the desired outcome. If you're writing a book review because your professor told you to, the professor's requirement provides one dimension of purpose. That requirement is not sufficient purpose to generate an effective review, however. You need to question yourself more closely: Why am I asked to write this? What does a book review do? You may then decide that you were asked to do the review to get you to read the book, think about it, and summarize the main points of the book and their relationship to the course.

Very likely all the biology writing you will ever do will have as its general purpose either to explain or to persuade. A further classification of kinds of explanation or exposition, as it is often called, will be useful.

- **Definition.** Used to answer the question, "What is it?" of your subject. *Example:* "What is ethology?"
- **Classification.** Used to answer the question, "What is the pattern?" of your subject. *Example:* "What are the various types of muscle tissue?"
- **Comparison and Contrast.** Used to answer the question, "What is it like or unlike?" of your subject. *Example:* "What are the similarities and differences between prokaryotic and eukaryotic cells?"
- **Analysis.** Used to answer the question, "What are the relationships among the parts?" of your subject. *Examples:* "What are

the roles of DNA, RNA, and ribosomes in protein synthesis?"

- **Argumentation.** Used to answer the question, "Can you prove it?" of your subject when your purpose combines explanation and persuasion. *Example:* "Should the federal government regulate emissions from industrial plants?"

WRITING 1.2: DEFINING PURPOSE. Begin with the broad subject, the eukaryotic cell. Then ask of that subject the questions listed above: for instance, "what is the basic cell structure?" Write your responses. The idea of this exercise is to learn how these purposes can help you sharpen a subject and focus your writing.

Decision 3: Audience

When you speak, you always speak to someone. That someone is your **audience.** It may be just one person, a group of friends, or your entire class. You know who the listener is and can see and hear reactions; you can tailor your talk to that audience, even modifying it according to the responses you get. When you write, however, your audience is unseen and perhaps even unknown. If it doesn't understand something you write, it cannot ask you to explain yourself. These differences between the speaker-listener relationship and the writer-reader relationship point up the importance of the writer's decisions about audience.

Just as you need to define the purpose of a piece of writing, you also need to define the audience. Let's take an example: the chairman of the biology department asks you to write a short essay to be titled, "Why Major in Biology?" If you agree to do the job, what additional information will you need? Certainly you need to know who is going to read this — what is your audience? The professor tells you that your essay will appear in a brochure to be distributed to students. Ah, but which students? Entering freshmen undecided about a major? High school seniors shopping around for a college? Upperclassmen considering a change of major? Although the **subject** seems to remain the same (reasons to major in biology) the requirements of these three audiences are somewhat different.

You can write much more persuasively if you can define your audience precisely.

What exactly does the writer need to know about the audience for a particular piece of writing? Although the answers to that question depend partly on the **purpose** of the writing, here are some useful questions to ask:

What is the gender of the audience? the age? job? income?

What is the audience's educational level? religion?

What does the audience already know about the subject?

What are its expectations likely to be? its attitudes?

What other special needs of this audience should you take into account?

Only by raising these questions will you discover which ones you need to concern yourself with as you plan a piece. By asking these questions about the audience of "Why Major in Biology?" you get a much better idea of how to direct that writing.

What about defining the audience of the writing you do in this course? Because student writers often have a great deal of trouble with that decision, you should know the options.

1. You can write for the professor, the most common way of defining the audience of a student paper. The problem with addressing this audience lies in your defining exactly who your professor is and what his or her expectations are.
2. You can write for the entire class, students and professor. If you do, you are likely to be less pompous and more direct than you would be addressing the professor alone. Why is that? How else might your work differ?
3. You can write for yourself, as you might in a journal (see Chapter 2). Practically every writer-in-training would benefit from producing more of this writer-based prose.
4. You can write for a specified audience, such as a medical school admissions board, people opposed to the construction of a regional power plant, or your congressional representative.
5. You can write for scholars in the field, as if you were composing an article for a professional journal. The best way to get a

sense of this audience's expectations is to read several articles in biological journals. See Chapter 7, Principles of Research, for suggestions.

WRITING 1.3: DEFINING AN AUDIENCE. Choose a topic that you are currently studying. Write an explanation of the ways that you might adapt your treatment of that topic for each of the five kinds of audiences listed above. Be specific about audience expectations and needs and about adjustments you might make.

Decision 4: Voice

"Who am I as I write this review or essay or whatever?" Although the question may seem silly, it really isn't. Your **voice** is the character, personality, and attitudes you project toward your subject, toward your purpose, and toward your audience. It is the "you" that you have deliberately chosen to express **on this occasion.**

Your writing voice, like your speaking voice, should be appropriate for the situation in which you find yourself or which you define. You don't use the same voice with your parents that you use with your best friend. Because we have so much experience speaking, we adopt the appropriate voice almost without thinking about it.

When we begin to write, however, we need to confront the choices consciously and to weigh a number of complex factors. Consider this variety of possible attitudes affecting your voice:

Subject: treat it seriously, lightly, humorously, reverently?

Purpose: praise, abuse, ask a favor, explain a process, encourage, persuade, complain?

Audience: peers, enemies, professor, lover, fellow vegetarians, other conservatives?

Occasion: formal, informal, ceremonious?

Clearly, these options are interdependent: that is, a writer probably wouldn't ask a favor in an abusive voice or complain to someone in authority in a humorous one. Your task as a writer is to match your voice to the occasion, subject, purpose, and audience.

Just how do you convey a voice, once you've selected one?

Word choice is one means. To the sensitive writer most so-called synonyms aren't equal. *Abdicate, resign, quit,* and *walk out* might mean roughly the same, but they aren't identical in meaning or in voice. Some words are simple and straightforward, others seem more formal. The distinction is made obvious in this pair of sentences that have the same meaning but different voices:

"She purveys Mollusca at the littoral area."
"She sells seashells at the sea shore."

Another stylistic element that conveys voice is **sentence structure.** A long, complex sentence might be appropriate for a relatively formal treatment of a serious subject for educated readers, whereas a series of short, declarative sentences could be more apt for informal treatment of the same subject for the same audience.

Even **punctuation** helps establish voice. A semi-colon, for example, is a fairly formal mark of punctuation. Writing in an informal voice, you would probably avoid that mark in favor of separating the clauses with a period. Dashes and exclamation marks generally have an informal effect. See Chapter 10 for other punctuation guidelines.

WRITING 1.4: DEFINING VOICE. Explain what voice you think would be appropriate for each of your responses in Writing 1.3. Be specific and relate your decision to concerns of subject and purpose.

Voice, audience, purpose, and subject — these are the key prewriting decisions. Making knowledgeable choices about each — that is your job as writer.

[2] *Keeping a Journal*

TOBY FULWILER

PREVIEW: *Writing in a journal will help you understand biology and the world around you. In this chapter you will learn what a journal is and how to use one.*

Why keep a journal?
What is a journal?
Characteristics of journals
Suggestions for your journal
What to write
Field notebooks

10–31 The thing that I find fascinating about insects is that everything is segmented. The legs are segmented into 4 parts . . . , the head is segmented into 5, the body into 13, the wing — God only knows! Now why would it be beneficial to the organism to have so many joints and segments? Why haven't a lot of these structures fused together over the years to more simplistic structures? I think the answer lies in the fact that insects have an exoskeleton. Therefore it is like turning a human being inside out. Now look who is more complicated and segmented!

—Student Journal Entry[1]

Keeping a journal can help you learn subjects like biology, botany, and zoology. The journal is a place to record observations, speculate, raise questions, and figure things out — as the student does about insect skeletons above. Journals have been used for such pur-

[1]All journal entries in this chapter were written in biology courses and are used with permission of the authors. No attempt has been made to correct errors or to improve style.

poses for a long time but only recently have they become widely used in college. Most serious thinkers, writers, scientists, artists, philosophers, and teachers have kept something like a journal in which to capture their thoughts. St. Augustine and Jean-Jacques Rousseau based their "confessions" on journals. Most of our Founding Fathers kept journals, as did authors such as Ralph Waldo Emerson and Henry David Thoreau, whose natural observations led to great literature. The major thinkers of our time — Darwin, Freud, and Einstein — recorded their questions and tentative answers in journals.

Samuel Pepys, the writer, called his journal a "diary." Edward Weston, the photographer, called his a "daybook." Albert Camus, the philosopher, simply a "notebook." Still others have called them "logs," or "commonplace books." Of course, it doesn't matter what this record of daily thought is called. What matters is that we understand why it can be useful and how it works.

If you have never kept a journal before, you might have some questions: What, exactly, is a journal? What does one look like? Why should I keep a journal? Aren't they something kept by poets rather than scientists? If I do keep one, what and when should I write in it? Above all, what can it do for me in this class? How can it possibly help me learn more about the life sciences? Let's look at some answers.

WHY KEEP A JOURNAL?

The act of writing helps people understand things better. Formal research as well as our own experience as students and teachers demonstrates this truth. If you are a student of the natural sciences and you write about theories, data, issues, and problems that you are studying, you will begin to sort out those theories, data, issues, and problems more clearly. Any assignment can be made richer by reflecting about it **to yourself** in your journal or notebook: What do I care about? What do I know? What don't I know? What do I want to know? What have I forgotten that I might remember if I wrote about it?

Writing helps you sort out and retrieve all sorts of information, ideas, and impressions already existing somewhere in your head. Notice what happens when you write letters to friends — how often you begin writing with one thing on your mind and then surprise yourself by bringing up all sorts of other matters. The same thing often happens when you write from an outline: you actually start digressing and going somewhere you never intended. And you like where you have gone and so need to adjust the outline accordingly. That's one of the remarkable powers of written language; it doesn't just reflect or communicate your thinking, it actually *leads* it! In other words, writing is a powerful mode of thinking.

Sometimes writing tells you flatly that you can't go where you thought you could! "Hmm, I thought I could explain why giraffes have such long necks — but if you don't allow for acquired characteristics, it's pretty tough. I wonder how Darwin handled that?" And so as you try to explain to yourself, you see the holes and recognize that you need more information. Learning when you're about to step onto thin ice can be a real survival skill — better to find that out in a private journal entry than in a public examination. Then you have time to do something about it: read more, research more, ask more questions, or whatever. Writing in journals about what you don't know is one of the best ways to start knowing.

Your journal will be both a place and a tool for thinking: use it to monitor class progress, to write daily plans, to rehearse for class discussion, to practice for examinations, and as a seedbed from which to generate research and term papers. Learn to trust that it will do that. Notebooks can be turned into journals when writers speculate on the meaning of someone else's information and ideas. Personal reflections about natural history can help you identify with, and perhaps make sense of, the otherwise distant and confusing past. Trial hypotheses about animal behavior might find first articulation in this same journal. Continued writing about theoretical ideas can develop those ideas into full-fledged research designs.

WHAT IS A JOURNAL?

I can give you an easy explanation first: journals assigned in class are essentially one part diary and one part course notebook. But a journal is also distinctly different from each of these. Diaries record the private thoughts and experiences of the writer. Class notebooks record the public thoughts and the presentations of the teacher. The journal is somewhere between the two. Like the diary, the journal is written in the first person ("I") about ideas important to the writer, but like the class notebook, the journal focuses on academic subjects the writer needs to learn more about. You could represent the journal this way:

Diary →	Journal ←	Class Notebook
("I")	("I/it")	("it")

Journals may be focused narrowly on the subject matter of a biology course or broadly on the whole range of your academic and personal experience. Each journal entry is a deliberate exercise in expansion: "How far can I take this idea? How accurately can I describe or explain it? How can I make it make sense to me?" The journal encourages you to become conscious, through language, of what is happening around you, both personally and academically.

CHARACTERISTICS OF JOURNALS

What's unique about journals is that they convey thought trapped in time — like moths preserved in amber. They offer an organizational pattern quite different from that of more traditional assignments. "Chronology" rather than "theme" provides the unity and sets the journal apart from other academic compositions. But while single journal entries are locked together in time, the collection as a whole may, in fact, transcend time to reveal more complex, often lucid, patterns of growth, development, and understanding. Unlike formal papers, journals carry with them all the time-bound fragments of thought since discarded, modified, or forgotten. Readers of journals, whether the writer or a teacher, get lots of chaff along with the wheat — and find nourishment there as well.

Language, too, sets journals apart. Some of the characteristics of good journal writing may run directly counter to traditional notions about appropriate academic writing. (We'll look at some actual samples shortly.) Journals may be full of sentence fragments, digressions, dashes instead of semi-colons, frequent references to oneself ("I"), misspellings, shorthand, doodles, sloppy handwriting, self-doubt, and all sorts of unexplained private references and notations. Both distracting and enlightening, these features occur in journals for different reasons than they occur in more formal writing. Journal writers must feel free to use their most comfortable, fast, close-at-hand style at all times. As a result, good journal writing is usually more fun to read — more like personal letters — than more carefully-crafted academic prose. The more we trust the value of our own informal voice, the more we will use it both to generate and to communicate ideas.

You should use your journal to experiment and play with language. Your sense of humor, normally suppressed in academic writing, is at home here. A student in a summer course wrote in her journal:

> We started this week talking about the dance of the bees. I was so nervous that I almost broke out in hives . . . . We were given a swarm of information.

We don't assign them for this reason, but we think journals are one place where writers can have fun if they feel like it. The most important thing is to write often and regularly on a wide variety of topics, to take some risks with form, style, and voice. Notice, for instance, how writing in the early morning differs from writing late at night. Experience how writing at the same time every day, regardless of inclination or mood, often produces surprising results. Above all else, your journal is a place where you can be honest with yourself (and your teacher), so write in the language that comes easiest to you. Here, for example, is a biology student's response to an in-class journal write, preceding a lecture on fungi:

> Actually I really don't know too much about fungi. When I think of fungi, I think of Athlete's Foot and a lot of itching. That's all I really know about fungi. I am embarrassed since I am a junior in biology. Fungi is small -- is it a one cell organism? It can't be too complex.

SUGGESTIONS FOR YOUR JOURNAL

The following list provides ideas for starting and keeping academic journals in virtually any subject area in college. But remember, these are just suggestions, not commandments. In truth, journals can look like and be anything you and your teacher wish.

1. Buy a small (7″ × 10″) looseleaf notebook.
2. Divide it into two sections: academic and personal (teacher will collect only the academic section periodically).
3. Date each entry; include time of day.
4. Write in your most comfortable, informal style.
5. Write daily if possible.
6. Write long entries, a new full page each time.
7. Write a personal reaction to every reading and class.
8. Freewrite (fast, without stopping) when you think you have nothing to say.
9. Collect quotes, clippings, scraps of interest.
10. At the end of the term, add table of contents and introduction.

WHAT TO WRITE

Journals are capable of containing any or all modes of symbolic thought which can be written or diagrammed. However they are especially useful for encouraging the very modes of thought most valued in the academic community. The following suggestions may give you some ideas for things to try out in your journal:

1. Observation. Use the journal to record, in your own language, what you see. The simplest observations are sensory experiences, primarily visual, but also aural, tactile, and the like:

```
9-4-85

Tree ident. #3

Leaves compound with pinnate leaflets and small
serrations. Bark smooth but cracked with upper bark
a whitish-grey color. Seven leaflets per leaf.
Leaflets are whitish beneath. Twig round, short
stems on leaflets. Leaves and leaflets opposite.
```

More complex observations try to capture whole experiences or events. And sometimes it's a good idea to record problems with observation, as we see in the following entry made by a student learning to identify trees with a taxonomic key:

> One tree in particular gave us a hard time and that was #2 on the map. We finally figured out it was a honey locust. We had to key it many times and each time we came to a dead end or got lost in the taxonomy key . . . . We found that the keys have a whole new vocabulary of their own.

In any case, the secret of good observation is being there, finding words to capture what you witness, and being able to experience it again when you see it recorded. In all science, observation is a crucial means of collecting data; journals can help you both collect and think about what you collect. Look for details, examples, measurements, and analogies. Use descriptive language including color, texture, size, shape, and movement.

2. Speculation. Use your journal to wonder "What if?" Speculation, in fact, is the essence of good journals, perhaps the very reason for their existence and importance. Journals allow writers to speculate freely, without fear of penalty — bad speculation and good, silly as well as productive because the bad often clears the way for the good and the silly sometimes suggests the serious. Use your journal to think hard about whatever possibilities — no penalties here for free thinking. Here is the first part of a page-long entry written by a student answering, thinking, wondering about his own observations:

> 1-11 This write concerns the fringe zone at the edge of the woods characterized by coarse grasses and shrub trees. This zone is a haven for insects and terrestrial organisms. In the particular fringe I'm observing a large amount of sumac is growing; why this is I have no idea. However I've also noticed growth of sumac in other areas where habitat is the same, yet rarely have I seen sumac grow in Dense Forest Areas. Despite plentiful sunlight the growth seems to be all small. Perhaps this is because the fringe I'm examining happens to be the

```
forest before it grows up. For example a grove of
trees spreads its seeds . . . .
```

This writer records what he witnesses in the fringe zone (sumacs) and why it seems to be that way ("the forest before it grows up"). In this case the writing guarantees that a record will be kept, but more importantly, the act of writing itself ensures that casual speculation takes concrete and extended shape.

3. Questions. Express your curiosity in writing; good thinkers ask lots of questions, perhaps more than they are able to answer. Questions indicate that something is happening — that there may be some disequilibrium or uncertainty in your mind, and that you are willing to explore it through language. Again, the ability to *see* one's questions certainly helps one sharpen, clarify, and understand them better. Sometimes writers use journals to record their doubts and uncertainties — one of the few places in the academic world where such frank admissions of ignorance have a place. (It may be all right to admit orally, after class, that you don't know an answer or understand something; it is something else altogether to admit it on an examination or formal essay.) In the journal one writes about what one does not know as well as what one does. Another name for a journal? A doubt book. Don't be afraid to write "What's that supposed to mean?" and "I just don't get this." In fact, in journals, it's as important to ask such questions as to answer them. Look at these questions jotted in a journal by another student trying to figure out what fungi are:

```
9-12  Is it considered a medium between plants and
animals? What exactly is its definition and de-
scription? Other than decomposition, what is its
function? Under what conditions does fungus grow?
```

Another student who, more than coincidentally is writing while lying on the lawn outside his dormitory, uses his journal to ask cosmic questions about insects:

```
9-16  If there ever was a nuclear war, what life
would survive? Would insects repopulate the world?
Would there be a new age of De-Evolution?
```

The journal is the natural place to keep a record of such queries, either for your own sake because you're intellectually curious or because such a record could help you raise questions in class or discover the subject for a research study.

4. Awareness. Learn who you are, record where you are, think about where you want to go. Be conscious of yourself as a learner, thinker, or writer. Self-awareness is a necessary precondition to both higher-order reasoning and mature social interactions. Journals are places where writers can actually monitor and witness the evolution of this process. You can encourage yourself to become more aware by asking lots of questions and trying out lots of answers: "What am I learning in here? What do I remember about today's lecture? About the assigned reading? What has any of this got to do with reality? With me? Why do I want to be a scientist, anyway? Or to get this college degree?"

5. Connections. Use the journal to make the study of the natural world (or any academic subject) relevant to everything else in your life — or try to. Can you make connections? Force connections? To other courses or other events in your life? Journals encourage such connecting because no one is insisting that writers stick to one organized, well-documented subject. Connections can be loose or tight, tangential or direct. The point is they are connections made by the writer (you), not somebody else. Digressions are also connections. They indicate that something is happening to trigger your memory, to bring forth information and ideas stored in your long-term memory. In journals, value them.

Here is a sample journal entry written by a student attending college in Michigan's upper peninsula; the connections are obvious:

> 10/10 I don't know if I'm just over-reacting to my Conservation class or not, but lately I've become suspicious of the air, water, and food around me. First we're taught about water pollution, and I find out that the Portage Canal merrily flowing right in front of my house is unfit for human contact because of the sewage treatment plant and how it overflows with every hard rain. Worse yet, I'm told raw sewage flows next to Bridge Street. I used to admire Douglass Houghton Falls for its natural beauty, now all I think of is, "That's raw untreated sewage flowing there."
>
> Our next topic was air pollution. Today I was informed that the rain here in the U.P. has acid

levels ten times what it should, thanks to sulfur oxide pollution originating in Minneapolis and Duluth. . . .

Is there any escaping this all encompassing wave of pollution? I had thought the Copper Country was a refuge from the poisonous fact of pollution, but I guess it's not just Detroit's problem anymore.

As I write these words, in countless places around the globe, old Mother Nature is being raped in the foulest way. I get the feeling someday she'll retaliate and we'll deserve it. Every bit of it.

It really doesn't matter in which class or at what time this writing occurred. The value is in having some place to record how one's school learning relates to the rest of one's life — in this case, to the writer's own backyard.

6. Dialogue. Talk to your teacher through the journal. Have a conversation, find out some things about each other — things perhaps too tangential or personal for class, but which build relationships all the same. When journals are assigned by instructors in academic settings, there is an explicit contract between student and teacher that entries related to thinking about biology will be shared. Consider this journal as "dialogical." Do not expect absolute, complete candor of each other — that's unrealistic anywhere anyhow. But journals can help you learn more about each other as co-learners if you share entries from time to time, either out loud in class or privately through written responses in the journal itself.

A Student's Journal Entry:

26 Nov. 85. The reptiles at one time were the most numerous animals on the planet. They evolved from an ancestral amphibian through the ability to live on land without having to return to the water to reproduce. They had a thick layer of skin which prevented water loss and were scaled. They were generally predatory animals. Were the dinosaurs scaly? Were they endotherms? Why was everything so large in comparison to today?

A Professor's Response:

- It appears that some of the dinosaurs *were* scaly. For others there is no evidence.
- They were *not* all predatory. Only a small % were.
- They were "so big" apparently as an unexplained trend of evolution towards largeness preceding extinction. Stephen J. Gould has a nice article on this.
- Whether they were endotherms (warm-blooded) is a hotly-debated topic today.

7. **Information.** Collect and comment on everything you can find that relates to the life sciences. Ironically, in a journal the straight factual information may seem like the least interesting material you collect; usually it serves more as record than anything else. A former student called the pages in which he recorded lecture notes "Cliff Notes stuff" and wished it were in his class notebook, not his journal. However, such references — especially when connected with some personal reaction — supply writers with valuable insights about otherwise rather distant material. You might even create a special section of your journal where you collect ideas encountered in your laboratory work, field trips, or readings. Label it "Topics for Further Investigation." Such a list can provide ideas for research and study in the future. Your journal is a natural place to keep them.

8. **Revision.** Consider your journal as a repository for scraps of thought for later development or revision. Journals are also places in which to re-think previous ideas. Try looking back in your journal and see if you can find places where you have since changed your mind on a subject written about earlier. Then write about what you now think and why you changed your mind. Anne Berthoff, a professor at the University of Massachusetts, advocates what she calls a "double-entry journal," in which writers return periodically to reflect upon previous entries. To keep a double-entry journal, make initial entries only on the right-hand pages. Leave left-hand pages blank for later use. This is a way to build opportunities for revision into the journal itself.

At other times, consider the journal as a place in which to start

a.

The purpose of this study was to investigate the increase of CO_2 on conifer growth; ~~in~~ and suggest both ~~the~~ regional and ~~effects and~~ long term effects. The ~~inv expe~~ investigation was conducted by studying tree ring samples, ~~studying temperat~~ seasonal temperature readings and reviewing and comparing previous pertaining data. The results suggest that ... at higher elevations ... photosynthesis

b.

The purpose of this study was to investigate the increase of CO_2 on conifer growth indicating both regional and long term effects. The investigation was conducted by studying tree ring samples, ~~season~~ recording seasonal temperatures, and reviewing and comparing previous pertaining data. The results suggest that at higher elevations, ~~the decrease of~~ photosynthesis is proportional to the decrease of carbon dioxide

c.

"Increasing Atmospheric Carbon Dioxide: Tree Ring Evidence for Growth Enhancement in Natural Vegetation.

The purpose of this study was to investigate the increase of carbon dioxide on conifer growth, indicating both regional and long term effects. The investigation was conducted by studying tree ring

d.

The purpose of this study was to investigate the increase of carbon dioxide on conifer growth, indicating both regional and long term effects. The investigation was conducted by studying tree ring samples, recording seasonal temperatures, and reviewing and comparing previous pertaining data. The results suggest that at higher elevations, photosynthesis is proportional to the amount of carbon dioxide partial pressure. Therefore, the results suggest that the increase of carbon dioxide has increased the growth rate of trees in higher elevation areas.

Figure 2.1 *Sequential entries in a student journal. a. First draft. b. Second draft. c. Third draft. d. Final product.*

formal papers — to make several starts until one idea begins to develop a life of its own. Then go with that one as far as you can until it busts loose from your journal altogether. Thus your journal can be a great place for first drafts to originate. Karen Brown used her journal this way to begin writing an abstract of a scientific paper. Compare her three journal drafts and the final copy reproduced in Figure 2.1 opposite.

9. Problem Posing and Solving. Use your journal to pose as well as solve problems. Don't make the posing something only teachers and experts do. Whether the problem is posed well, or whether the solution actually works, matters little. (If the problems are consistently ill-defined and the solutions always off base, that does matter, but here, the journal will be invaluable in another way, as an early clue to where you are really having trouble.) According to Brazilian educator Paulo Friere, individuals must articulate problems in their own language in order to experience authentic growth. Journals are, perhaps, the best place in the academic world in which to do that. Evidence of posing and solving problems — whether literary, social, scientific, or mechanical — suggests that you're alive, thoughtful, and perhaps even committed. In the following entry we see a student reflecting on a previous entry about the Lamarckian theory of evolution, then actually posing his own quite specific questions which could be the basis of further study:

> 10-17 I just want to elaborate on one section of Lamarck. How the Giraffe got its long neck from stretching up to get the trees. However, even Lamarck could not have believed that the neck became 30 ft long . . . . Therefore my question is What did the Giraffe eat before his neck became long so he could eat the trees? Why didn't the population of Giraffes starve to death? Well, maybe they ate something else! Well, if that's the case, then their necks wouldn't have developed to eat the trees . . . . Maybe the trees were once short and they acquired the ability to get tall from having to be eaten by the Giraffes?

Professor's Response to Journal Entry:

> Actually this is a nice start to a paper on the evolution of thought on natural selection. You could address each point. All of these questions have been asked and answered, so the literature is there.

The point, of course, is not so much that the writer find a sophisticated or authoritative answer — he doesn't — but that he work toward an answer to a problem that has suddenly confronted him. This entry, like some of the other samples we have looked at, illustrates several modes of thought taking place in rapid succession: asking questions, expressing doubt, analyzing a problem, and trying out a solution. In that sense, it is also an example of the last suggestion I will make here: synthesis.

10. Synthesis. One of the best and most practical activities to do with your journal is to synthesize, daily and weekly, what's going on in your science study. "How does this lecture relate to the last one? What do I expect next time? How does class discussion relate to the stated objectives on the syllabus?" Your written answers to any of these questions can easily generate comments to share with both class and teacher. If you can take even 5 minutes at the end of each lab or lecutre — or stay in your seat 5 minutes after class — you can catch perceptions and connections that will otherwise escape you as you run off to another class, lunch, or a quick snooze back at your room. Journals invite you to put together what you learn.

FIELD NOTEBOOKS

In the natural sciences your instructor may ask you to keep a very particular kind of journal called a **field notebook.** Such a document is actually a cross between a journal and a notebook, written with the deliberate attempt (1) to record as objectively as possible all data from whatever fauna, flora, field, or stream one is observing and (2) to keep a running record of speculations about those observations. In such notebooks dates, times, places, temperatures, site conditions are crucial for later work in laboratory, study, or library. In other words, a field notebook does very much the same things as a journal, but is more focused in its intention. If you are asked to keep one, most of the suggestions provided in this chapter will apply. Here, for example is a summary entry in a field notebook, recorded after the writer returned from an outing:

> 9-18-85 Today went to the reservoir, pond located in Winooski, VT on the nature trail at SMC. Collected a water scorpion, several surface insects, and many nymphs. Caught the insects by using net just along the edge of the water and bringing it into the grass on the edge. The specimens were taken out of the net and placed in a jar containing alcohol. The alcohol killed the specimens, and left them in their natural shape. Upon returning to the lab room we separated the ones we knew and changed the alcohol in the others. We will later identify them and put each specimen in its own alcohol filled vial.

A Few Last Words

If you let journal writing work for you in some of the ways suggested in this chapter, you'll gradually learn to be both a better learner and a better writer. Journals aren't magic. But the practice of daily speculative writing will exercise your mind in much the same way that running or swimming exercises your body. The practice of writing to oneself can become a useful regular habit. Try fifteen minutes each morning with coffee, twenty minutes each evening before homework, or even ten minutes before bed. You will find writing is easier and easier and, in time, may find it a mentally restful activity — the one time in a busy schedule to put your life in order. And at the end of the term or after you graduate you'll appreciate this marvelous written record of your thoughts, beliefs, problems, solutions, and dreams. The nice thing about the journal is that it represents a powerful process while you keep it and, at the same time, it results in a wonderfully personal product.

[3] *Learning Biology Through Writing*

PREVIEW: *The problem of getting started and the reading/writing connection are the topics of this chapter. For many people getting started is the hardest part of writing. We'll show you some techniques that help. We'll also show you how to write about your reading.*

- *Techniques for getting started*
  - *Try focused free writing*
  - *Talk it out*
  - *Write a table of contents*
  - *Establish a rhythm*
- *Reading for main ideas, writing to summarize*
  - *Reading for main ideas*
  - *Writing to summarize*
    - *Summary*
    - *Abstract*
    - *Book review*

Science is organized knowledge.

—Herbert Spencer

Earlier in this book we claimed that writing is an invaluable way to learn biology. Writing helps you discover what you know and what you think. Writing leads your thinking toward new insights. And if science is organized knowledge as Spencer claims, writing is the means to that organization.

If you've been keeping a journal, following the suggestions in the last chapter, you probably are beginning to agree — writing does help you to ask questions, make connections, pose and solve problems, even synthesize and create new knowledge. Your journal is a great personal tool for thinking about the living world. But what happens when you need to go public? How do you meet those demands made by professor or boss to write to a larger audience than yourself?

You'll find some of the answers in this chapter, beginning with some proven strategies for getting started on any kind of writing assignment for any course. Then we take up those skills so critical to the student of biology: reading for main ideas and writing to summarize. They're not simple answers. Writing isn't easy. But if you work at it, you'll develop abilities that will take you through your college years and beyond — through graduate or professional school, into your first job, and throughout your life.

TECHNIQUES FOR GETTING STARTED

It's the first day of school. Along with the excitement of beginning a new semester — new courses, new classmates and new professors — comes the anticipation of new assignments. Some assignments can appear fairly intimidating, others stimulating. But you begin the semester with a new resolution: you're not going to let your assignments slide. No more waiting until the last minute the way you usually do.

But despite the best of intentions, it often seems difficult to get going on a writing assignment. You may feel that all you have is a jumble of ideas and you don't know how to begin assembling them on paper. You decide to get another cup of coffee or play one more game of trivia. Or perhaps you feel that your mind is a blank — you don't have any ideas at all for the assignment. So you keep putting it off. Maybe inspiration will suddenly strike in the midst of a TV show.

Just how should you get started on a writing project? Is it best to make a detailed plan of what you wish to say before you begin to say it? Or should you jump right into the assignment, let your language flow, and worry about organization later? There is, of course, no correct way to begin a writing assignment. Different strategies work for different people. Let us suggest some techniques that have worked for us.

Try Focused Free Writing

It's quick! It's painless! No one will ever see it but you! You can ignore it if it doesn't seem to get you anywhere. Focused free writing involves plunking yourself down with pen and paper (or at your computer keyboard) and forcing yourself to write on your topic rapidly and without judgment for a limited period of time. "But I don't have anything to say! That's why I'm having trouble with this assignment!" Don't try to predict what you have to say. Don't worry about writing anything profound or even coherent. Focused free writing is a type of discovery draft, a technique to draw out your thinking. Don't try to plan what you're going to write. Don't worry about what will come out of it. Just put your pen to paper and make yourself write continuously whatever comes to mind. The only rule is to stay focused on the topic.

Consider this student's attempt to get started on an assignment using focused free writing.

> What a topic -- "could parasites be the origins of free living forms?" I just started this invertebrate course and already I'm over my head. What do I know about parasites? I know that they're things that live on other forms of life. What kinds of things? Ticks? Fleas? How could these be origins of free livers -- they are free livers! Maybe he means inside -- what are they called? Endo-parasites. What are those? Tapeworm must be one, and the liver fluke, and the blood forms -- the ones that give us malaria. Would one of these fit? What do you know -- there's more GB left in me than I thought. What else do I remember? Not much. BUT the library must have something on parasites. Maybe someone else has already answered this dumb question. I'll stop on my way to the BB game tonight and start a list of internal parasites. That's settled. Now how can I put all of that information to use. No sense reading any more than I have to if I

```
can figure out now what I need. Probably only the
internal parasites are important for this topic.
They seem simpler (less complex, I mean) and I
could try to explain what they would have to do to
become free living. Maybe even find examples of
free living forms that resemble parasites. That's
not a bad idea. I even think I can find examples of
some parasites that have partial free living stages
-- boy, that would help!
```

It's not surprising that this free writing starts off with no apparent direction. That is the nature of a free write. But look where it leads after only a few minutes. First, she begins to realize that she does know something about the topic. Even more importantly, she has decided what to look for, where to look, and how the material might be arranged. There is still much to be done, but she has a skeleton with which to work. A more apt comparison would be a few bones which will have to be arranged with other, yet to be discovered, bones into a skeleton. The skeleton will be the outline or table of contents which can then be fleshed out into the final paper.

That's the whole idea of free writing — getting started. No paleontologist ever found a bone without a lot of spade work, metaphoric if not actual. Sitting in camp, hoping for a discovery doesn't produce results. Sitting at your desk without actually writing is also nonproductive. Free writing has led other scientists to insights. It can do the same for you.

Talk It Out

Another strategy for discovering your own thoughts is to talk to someone about your subject. Discuss the project, try to explain your basic ideas. You may find it helpful to speak to your professors since they can often ask relevant questions and give direction to your thinking. But talking to a friend can be just as useful, if not more so. Those who know little or nothing about your topic may ask you precisely those basic questions which should lay the groundwork for your essay. The input from your listener is only part of the benefit of talking it out. The very process of structuring your language through speech will help you become more aware of what you know and don't know, where your logic is clear and where fuzzy, which ideas are relevant to your larger thesis. In fact, you should use this technique throughout the composing process. It's particularly helpful in working through points where you just can't find the right words and where a paper has lost its focus.

Write a Table of Contents

Shortly after talking through the general design of your essay, try composing a brief table of contents or outline of the major points you expect to cover. Some writers cringe at the idea of an outline because of all those associations they have with doing outlines in 7th grade English class. It seems like such a tedious, if not intimidating, process when you get hung up on the details of its form. One solution is not to write an outline, but a tentative table of contents just for your own use as you draft. Make it rough, make it loose, but jot down the main ideas in the order you think they should appear.

On the left of Figure 3.1 is an extremely rough table of contents

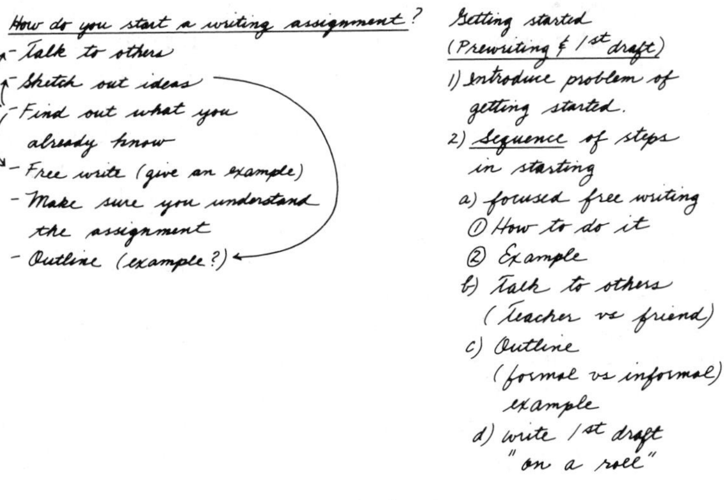

How do you start a writing assignment?
- Talk to others
- Sketch out ideas
- Find out what you already know
- Free write (give an example)
- Make sure you understand the assignment
- Outline (example?)

Getting started
(Prewriting & 1st draft)
1) Introduce problem of getting started.
2) Sequence of steps in starting
a) focused free writing
① How to do it
② Example
b) Talk to others
(Teacher vs friend)
c) Outline
(formal vs informal)
example
d) write 1st draft
"on a roll"

Figure 3.1 *Two drafts of contents.*

for the first part of this chapter. On the right is a subsequent revision of the original, written after the chapter was under way.

Our routines are always subject to dramatic changes and rearrangements. In the middle of writing we often get new ideas which reshape our original thinking. In fact, making a new outline after you've finished a first draft helps you check the logic of your writing.

Establish a Rhythm

In an essay entitled "A Writer's Discipline," Jacques Barzun (1971, p. 8) argues that you must "convince yourself that you are working in clay and not marble, on paper and not eternal bronze; let that first sentence be as stupid as it wishes. No one will rush out and print it as it stands. Just put it down; then another." It is crucial to establish a rhythm to your work. If you can't think of the precise word you need in a particular context, continue without it. If you can't think of an appropriate example where it's needed or are having trouble deciding between two possibilities, just make a note to yourself in the margin ("example" or "?") and get on with the drafting. Rule number one in writing and Las Vegas: never stop when you are on a roll.

Don't agonize over spelling or grammatical rules. These matters are better left for revision or, in the case of an essay exam, when you are ready to proofread your paper. During the composing process it helps to permit your language to flow as rapidly as possible. This is especially true in an essay exam when you are working against the clock. As you establish a momentum, you start to gain a sense of confidence. You begin to realize how much you actually know about your subject, to see ways of organizing that information, and, just as important, to recognize what issues you need to think about further. When you reread your first draft, you will undoubtedly discover elements of repetition and occasional moments of incoherence. But don't attempt to polish the material until you are satisfied with the overall structure. Why bother tinkering with a sentence before you're certain you want to keep it at all?

Establishing a rhythm means getting your thinking onto paper or PC disk, no matter how rough that thinking may be. Not even the most accomplished writers produce first drafts that say exactly what they intend.

Our guess is that many people would prefer to ignore all we've said above. It's tempting to try to write a first and final draft from scratch because the very idea of revising seems long and tedious. But think about the time you've wasted staring at a blank page. The processes of focused free writing, talking, outlining, drafting, and revising are often actually quicker and usually more productive than trying to compose the final product in one draft.

WRITING 3.1: GETTING STARTED. Develop the foundation of a brief (1-2 page) essay. Begin by choosing an interesting topic covered in your course, perhaps enlarging on a promising journal entry. Then use as many of the suggested techniques as you can. Try focused free writing. Talk it out with a friend. Then list the sequence of points you hope to cover and draft the essay. You should end up with a good draft of a potentially fine piece of work.

READING FOR MAIN IDEAS, WRITING TO SUMMARIZE

The connection between reading and writing is nowhere more evident than in the production of an abstract or summary. Whether you think of yourself as a reader writing or as a writer reading, first you must discover the thrust and main ideas of a journal article or book and then convey that information precisely and without wasted words.

Reading for Main Ideas

Reading for main ideas is not much different from the kind of reading you do when you study new material in a textbook. In fact, often it's the same kind of reading. The purpose is to identify and extract the gold from all that ore, in other words to find the thesis, or assertion, of the article or book chapter and the supporting points. But how do you locate those main ideas? To a large extent, the material you work with will determine the approach you take:

each article or book will differ in organization and presentation. But because most professional writers follow certain rules of organization, we can offer some basic principles of reading that are generally applicable. Applying these hints should help you read most material with understanding.

Perhaps the simplest, yet one of the most effective reading strategies, is to use the typographical signals in the text itself. Typographical signals are meant as signposts to guide the reader through the maze of the text, just as route numbering signs are intended to lead the driver through an unfamiliar city. Begin by thinking about the title of the article or book. Don't overlook the subtitle, if there is one. These features identify the subject and often the scope of the work. Chapter and section headings tell you what is to come. Within paragraphs look for words printed in **bold face** and *italic* type, indicating important terms or concepts. Supporting data is usually summarized in a graph or figure; if the significance of that data is not clear, look to the corresponding section of text.

Another reading strategy is to look for key ideas in the obvious places. If you are dealing with an entire book, look to find the thesis, scope, and method explained in the preface or introduction. For an article or book chapter, look for the thesis at the end of the first paragraph or the beginning of the second paragraph. Within the piece, main ideas are ordinarily where you would expect to find them — at the beginning (or, less commonly, at the end) of each paragraph.

Writing to Summarize

Imagine yourself in this situation: you have just spent two hours struggling through a complex article about recombinant DNA. As you read, you made notes in order to help yourself decipher the article and remember what it is about. When you arrive in class, your roommate, who barely made it through the first page of the article, pleads, "Quick, fill me in!" And you try. Then your professor enters the classroom and asks you to jot down the key issues in the article. Finally, as you leave the class, you are given the assignment of writing an abstract of the article for your next class meeting.

In each of these four instances you have a need to summarize the material from a journal article. However, the purpose and thus the form and content of each summary is a bit different. In this section we'll explain briefly how to write article summaries as well as a related form, the abstract, and how this process is shaped by your purposes.

Let's begin by distinguishing between abstracts and summaries. In general terms an abstract is a variety of summary, but the term *abstract* has taken on a special meaning in scholarly writing. The abstract in biology (and in the other sciences as well) has requirements of content and form that have been standardized virtually worldwide. An abstract is formal and impersonal; its format and length are prescribed by the conventions of the discipline. On the other hand, a general summary of a paper or book may be informal and personalized; its form, content, and length vary to suit the writer's needs. We'll look first at the summary, then at the abstract.

Summary

Just about the only rule that applies to writing a summary is this: present the basic ideas of the original without distortion. The summary's form and content are dictated by its intended purpose, subject, and audience. In the hypothetical example described earlier, you have several reasons for summarizing an article. You might write a summary in your journal to help clarify your thinking about the article. In this case, your summary might or might not be comprehensive and detailed. It might or might not be in a form that would be understood by another reader. This summary might meander back and forth between your personal experience and only one of the themes of the article. Or it might compare this and another article's perspective on a relatively small but significant issue. Alternatively, you might have summarized the article so that you could use it in a research paper you are working on, in which case you'd probably want to be more comprehensive and include all the key points made in the article.

The summary of the article you made for your roommate is likely to be much different. Your purpose here is to communicate a general overview of the information to someone who is unfamiliar with the material. This summary needs to teach another person new information rather than remind you of connections you have already considered. While less comprehensive than your written summary for exam preparation, it must be relatively detailed in the points you do make.

Your five-minute class exercise in jotting down the article's key issues for the professor is unique in yet other ways. Your purpose is to evaluate, that is weigh the significance of the issues raised. You must be comprehensive in terms of pointing out key issues, but you need not be particularly detailed on any. You assume that your audience (the professor) is well versed in the area, so your intent is to identify and integrate issues rather than to reiterate details or teach the information to a naive reader.

Because summaries can serve so many differing functions, you should identify your purposes from the outset. The most efficient and effective summary writing is highly sensitive to its intended purpose, audience, and subject matter.

The following student summary was submitted in a sophomore biology writing course. Can you tell the student's purpose and audience?

The author of "Plasma levels of zinc in protein-calorie malnutrition and after nutritional rehabilitation" (Nutrition Review, July 1983) clearly states his purpose in a concise sentence following the title: to show the reduction of plasma zinc in protein-calorie malnutrition and its restoration through suitable oral zinc supplementation. The author wrote this article, I believe, to present evidence that protein-calorie malnutrition is a contributing factor in acquired zinc deficiencies, to reveal the effects on the body of a zinc deficient diet, and to suggest a method of restoration of normal levels of plasma zinc through dietary supplementation.

The paper begins by stressing the importance of a zinc rich diet through the revelation of some effects caused by zinc deficiency. These include growth retardation, anorexia, diarrhea, and mood and temperament changes. Pregnant women and growing children are most vulnerable to acquisition of zinc deficiencies. Little statistical evidence or specific experimental data are offered to support these statements. I believe, however, the information is sufficient, for the paper's purpose was not to show the steps of degradation from a healthy person to a zinc-deficient one, but to correlate protein-calorie malnutrition and plasma levels of zinc.

The author does effectively show a correlation between zinc levels and malnutrition. The consequences of protein-calorie malnutrition parallel many of the effects of zinc deficiencies. Studies of malnourished children conducted in India, Nigeria, and Ibadan all indicated that an improper and/or insufficient diet results in reduced zinc levels. Data were presented showing specific zinc concentrations in malnourished children. I would agree with the paper's conclusion, for groups were studied over a wide geographical distribution and all evidence is in agreement.

The author devoted the final section of the

article to a discussion of the restoration of normal levels of zinc in individuals with zinc deficiencies. I believe this constitutes the strongest part of the paper. It re-emphasizes how vital zinc is to proper biological functioning but goes on to propose that the addition of extra amounts of zinc to rehabilitation food supplements can repair much of the damage incurred by a zinc deficiency. The author presented a problem, showed the effects and the extent of that problem, and most importantly offered a possible and effective solution.

The entire article was clear and concise. It presented the concern (plasma levels of zinc), correlated it to a probable cause (protein-calorie malnutrition), and offered a means of solving the problem (nutritional rehabilitation). The article provides informative background on the subject and can easily be read without extensive biological knowledge.

WRITING 3.2: WRITING A CRITICAL SUMMARY. Select an article from *BioScience*, *Science*, or a journal assigned by your professor and write a critical summary of it. Use the following questions to guide your reading and to organize a one- to two-page written summary.

1. Can you tell what the author's purpose was in writing the article?
2. If this article is the result of an experiment, why was the experiment undertaken?
3. Can you state the author's hypothesis for the experiment?

4. What assumptions were made in formulating the hypothesis?
5. Does the experimental design truly allow the testing of the hypothesis?
6. Do you agree with the author's conclusions?
7. Is the title appropriate for the contents of the article?
8. What do you think are the strengths and weaknesses of the article?
9. Consult the letters to the editor section of several subsequent issues of the journal. Summarize any letters treating your article.
10. Who would you recommend read this article?

Abstract

The purpose of an abstract is to provide a concise but comprehensive recapitulation of an article or book, an overview that will allow readers to decide whether the article treats information that is of interest to them. Abstracts are ordinarily used in two contexts. The abstract is the first section of any article appearing in a professional or scholarly journal. Abstracts are also used in indexing or information services (e.g. *Biology Digest*) where they are organized by subject, date, etc. to allow readers to systematically search for information of interest. The *BIOSIS Guide to Abstracts* (CBE 1983) defines an abstract as "a noncritical, informative digest of the significant content and conclusions of the primary source material. It is intended to be intelligible in itself, without reference to the paper, but not a substitute for it." As you may have already found from your own reading, a well-written abstract provides enough information to stand alone as a small version of the total paper. Abstracts are generally the last part of a paper to be written. In many ways they are a summary of a summary.

Do you need to worry about how to write an abstract only when you are ready to present your research for publication? Not at all. Composing an abstract serves an important function for the reader/writer: it forces you to analyze carefully the true essence of the material you're reading or writing. The significance of this process becomes clearer when you understand the specific form and content requirements of a formal abstract. These requirements vary slightly depending on the nature of the article you have to treat. Therefore, we'll consider each type of article and its abstract in turn.

A common kind of article in biology is the **research report,** a presentation of the methodology and findings of a scientist's research. An abstract for a research report should be no longer than 250 words — even if the work took five years to complete and the article is 25 pages long. The abstract is typed as a single paragraph. According to CBE (1983, p. 20) guidelines, it should include the following elements:

1. The objectives and purposes of the study.
2. The materials, methods, techniques, and apparatus and their use as well as new items and new applications of standard techniques and equipment.
3. Scientific and common names of organisms (if given in the article) with special emphasis on new taxa or new distribution records.
4. Specific drugs (generic names preferred) and biochemical compounds, including the manner of use and route of administration.
5. New theories, terminology, interpretations, or evaluations concisely stated.
6. New terms and special abbreviations and symbols defined.

An abstract should **never** include:

1. Any material not in the text.
2. Figures and tables or references to them.
3. Comments on how to follow specific methods or use the apparatus.
4. Literature references.

The following abstract from *Limnology and Oceanography* (Nov. 1985, p. 1240) meets all of these criteria.

The Horizonal Heterogeneity of Nitrogen Fixation in Lake Valencia, Venezuela[1]

SUZANNE N. LEVINE AND WILLIAM M. LEWIS, JR.
Department of Environmental, Population and Organismic Biology, University of Colorado, Boulder 80309

Abstract

Spatial and temporal variability of nitrogen fixation in Lake Valencia, Venezuela, were quantified on the basis of duplicate water samples collected from a depth of 0.5 m at 16 sites on 10 dates. The concentration of heterocysts in samples was determined and the samples were incubated with acetylene in situ. Two-way ANOVA was used to separate the variance associated with site (fixed spatial patchiness), date (temporal variation), the interaction between site and date (ephemeral spatial patchiness), and sampling error. The nitrogen fixers in Lake Valencia are arranged in large (40–200 km^2), ephemeral patches with distinctive fixation rates per heterocyst. Both variability in fixation per heterocyst and variability in heterocyst concentration contribute significantly to variation in fixation per unit volume of lake water, but the variability attributable to heterocyst abundance is greater. Spatial variation in fixation and heterocyst concentration exceeds temporal variation in these parameters, and the ephemeral component of patchiness is much greater than the fixed component.

A second kind of article written by biologists is the **review article,** which is either a review of the literature in a sub-field or the presentation of a principle of biology. An abstract of one of these articles is quite similar to that for a research report, but is likely to be a bit shorter — about 150 words. An abstract of a review article would not include a description of procedures, but would summarize the article and have a conclusion. The following abstract, from *The Quarterly Review of Biology* (1977, p. 371), illustrates this type, although it is somewhat longer than usually recommended.

[1]*Limnol. Oceanogr.*, 1985, 1240–45.

Aspects of Biological Exploitation[2]

MICHAEL L. ROSENZWEIG
Department of Ecology and Evolutionary Biology, University of Arizona, Tucson, Arizona 85721

Abstract

The population interaction termed biological exploitation includes what has formerly been called predation, as well as other interactions in which one population takes advantage of another (e.g, grazing, parasitism, Batesian mimicry). An instantaneous, deterministic theory using graphs of the properties of difficult and even unknown autonomous nonlinear differential equation systems has been developed to simplify greatly the task of understanding the dynamics of such systems and of predicting qualitative properties of their solutions. This review shows how the theory may be used to account for some of the observed dynamics of well-known laboratory systems including their oscillatory periods and neighborhood stability. It also extends the theory to cover situations where the predator prefers to attack weak or otherwise vulnerable victims. In this case, an upper limit is shown to be added to victim oscillations which can serve to promote the survival probability of the system despite the fact that it may diminish the system's ability to return to a steady state following perturbation. Finally, the theory is applied to the problem of management of a pest by biocides. An important result of earlier predation theory, which has tended to be discarded because of the oversimplifications in that work, is shown to hold in many of the more realistic situations described by graphical exploitation theory: addition of a biocide can actually increase the average population densities of the species one intended to attack. The principle at work here may well account for some modern instances of crop-pest population explosions, and could have helped to predict which pests were likely candidates for control by chemical means and which were not.

---

[2]*Quart. Rev. Biol.* 52:371 (1977).

As you can see, the abstract must be very brief but also comprehensive. That's where the difficulty lies. How can you put all that information into 150–250 words (a typed page or less)? It definitely is not easy. That's why writing an abstract provides such excellent practice in distilling, communicating, and preserving the essence of an article or project. Writing an abstract, whether based on your own research or someone else's, can crystallize your understanding and appreciation of the basic elements of a piece of scholarly writing.

Although the Council of Biological Editors (CBE 1983) treats abstracts, we have also found some excellent advice in what might seem an unlikely source: the American Psychological Association style sheet (APA 1983). APA's description is well worth keeping in mind. A good abstract is

1. **accurate.** The abstract must accurately reflect the paper. Not only should you avoid actual errors in representing the article, but you should also avoid shifting the emphasis. Put nothing in the abstract that isn't in the original.
2. **concise and specific.** Each sentence should pack as much information as possible. To be concise be specific.
3. **self-contained.** Your abstract should be able to stand on its own without reference to the original paper. You should define acronyms, abbreviations, and unique terms, and spell out the names of uncommon procedures. Paraphrase rather than quote.
4. **nonevaluative.** Simply report what's in the paper; the abstract is not the place to add your own thoughts and insights.
5. **coherent and readable.** Use clear prose and active verbs rather than passive voice ("analyses revealed" rather than "it was revealed by analyses"). Use past tense to refer to procedures used or variables manipulated in the study. Use present tense to refer to results that have continuing application and to conclusions drawn.

Elements of an Abstract

For a research report
1. the problem
2. the subjects
3. the procedure
4. the results
5. the conclusions, implications

For a review article
1. the topic
2. the thesis and scope
3. the main points
4. the conclusions, implications

WRITING 3.3: WRITING THE ABSTRACT. We've discussed the form, content, and characteristics of a good abstract. Is there a trick to putting it all together? Not really, but there is a process you can follow that will yield an effective abstract.

1. Read the article from *Science* (1985, pp. 1024–1025) that follows this assignment. Then review the elements of the **review abstract** above. Write a brief statement based on the article about each of these elements. Continually refer to the article to be sure that you are accurate.
2. Now review the list of information you've compiled and decide whether each point listed is truly essential. Remember the abstract is intended to represent, rather than fully describe.
3. Next try to put the abstracted information together into a logical series of complete sentences. Count the number of words to see how you're doing. Your target is 250 words.
4. Revise your draft, attempting to be more concise and specific. Go back to the article for confirmation of what you've written and to be sure you haven't distorted its meaning. You may have to repeat this step several times.
5. When your draft satisfies you, check it against the qualities of a good abstract explained earlier in this chapter.

What Makes Nerves Regenerate?[3]

DEBORAH M. BARNES

Damage to nerves often results in a permanent loss of function because the tissue fails to regenerate. This is a particular problem in the mammalian spinal cord and brain. But now it seems that the regenerative capacities of both the peripheral nervous system (PNS) and the central nervous system (CNS) are considerable, if the appropriate conditions are provided.

Research groups headed by Ioannis Yannas at the Massachusetts Institute of Technology (MIT) and Richard Sidman at the Children's Hospital, Harvard Medical School, have worked independently to devise strategies for enhancing the regeneration of peripheral nerves (those outside the spinal cord and brain). Each group has induced nerve fiber outgrowth over a distance of 15 mm or more by using tubes that allow the infiltration of nonneuronal cells whose presence is essential for extensive regeneration.

At the same time, Albert Aguayo and his colleagues of McGill University in Montreal, Quebec, have demonstrated that CNS axons have an extensive ability to regenerate in the presence of peripheral nerve grafts. Aguayo's recent work is with the eye, and it illustrates two important phenomena: damaged CNS fibers can regrow through a PNS graft over a distance greater than normal, and some of the regenerated axons regain functional properties.

Still another approach is to identify the cell types and chemical substances in the peripheral nervous system that support regeneration. This is

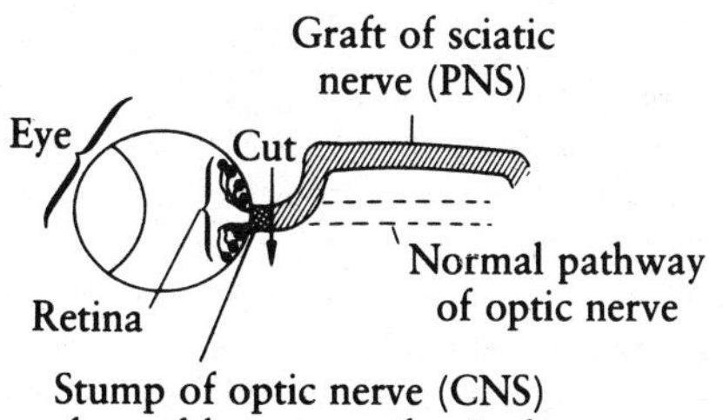

Cutaway view of severed sciatic nerve regenerating through a synthetic tube used by Yannas and his colleagues (used with permission of I. Yannas).

[3]From *Science* 230:1024–1025. Nov. 1985. Copyright 1985 by the AAAS. Used with permission.

a direction explored independently by Hans Thoenen of the Max Planck Institute of Psychiatry in Martinsreid, West Germany, Paul Patterson at California Institute of Technology in Pasadena, Randy Pittman of the University of Pennsylvania School of Medicine in Philadelphia, and also by Sidman's group at Harvard. Thoenen's most recent findings indicate that cells from the PNS form a "permissive" substrate for nerve regeneration, but CNS cells form a "non-permissive" substrate. Patterson and Pittman have identified a potential role for protease enzymes and their inhibitors in axon elongation.

Investigators generally agree that severed peripheral nerves, such as the sciatic, which leaves the spinal cord to innervate the leg, will sprout to form the beginnings of new fibers. But some assistance is required to guide the direction of sprouting, if the axons are to make specific functional reconnections with their normal target tissues. One way of directing axonal regrowth is to provide a tube through which cut fibers can grow across a gap. In 1982, Goran Lundborg of Göteborg University in Sweden, Silvio Varon of the University of California at San Diego, and their co-workers described the use of a silicone tube that allowed rat sciatic nerve regeneration across a 10-mm gap (*l*).

Yannas, a polymer chemist at MIT, and Jerry Silver, a neurobiologist at Case Western University in Ohio, induced more extensive regeneration "by restructuring the space inside the guiding tube." Also involved in the project were Dennis Orgill of MIT, Thor Norregaard and Nicholas Zervas of Massachusetts General Hospital, and William Schoene at the Brigham and Women's Hospital in Boston. In a presentation to the American Chemical Society, held in Chicago in September, Yannas reported rat sciatic nerve regeneration across a 15-mm gap using a silicone tube packed with a protein, collagen, and a glycosaminoglycan (GAG) polysaccharide, chondroitin-6-sulfate. These materials are cross-linked to form a porous network that is degradable by enzymes at rates that can be controlled during preparation, although the silicone tube itself is not biodegradable.

Yannas originally developed the polymer in 1975 to promote skin regrowth over an area that had been severely burned. His primary objective with both nerve and skin was the same. He wanted to find a material that was capable of enhancing regeneration without using an autograft and could also be manufactured easily. He believes his most important finding is that almost identical materials stimulate the growth of both nerve and skin, tissues that are very different. The regenerated nerve tissue looks healthy: it is highly vascularized and about 20 percent of the axonal fibers are surrounded by a fatty myelin sheath produced by Schwann cells that have migrated into the tube.

Meanwhile Roger Madison, Ciro Da Silva, and Pieter Dikkes in Sidman's department induced sciatic nerve regeneration over a 20-mm gap through tubes lined with either a collagen matrix or a laminin-containing gel. Their work was described at the Annual Meeting of the Society for Neuroscience*.

Laminin is a glycoprotein associated with the basement membrane (basal lamina) of Schwann cells and is part of the normal substrate for nerve fiber outgrowth during development. Sidman, Earl Henry, and their colleagues at Harvard, in collaboration with Tri-Ho Chin of Allied Corporation, have developed nontoxic nerve guide tubes. The tubes are filled with gels rich in laminin and other extracellular matrix materials. These materials stimulate greater numbers of regenerating axons (many of which become myelinated), faster outgrowth of fibers, and the ability of regenerating axons to span longer gaps (*2, 3*).

Sidman also notes that the presence of the distal stump at one end of the tube is critical for regeneration. He believes that the "tube isolates the local milieu," thus establishing conditions that encourage fiber outgrowth. The "milieu" consists of all the tissues invading the tube, including nerve fibers, fibroblasts, blood vessels, macrophages, and Schwann cells, as well as any growth-promoting factors they may secrete and any extracellular matrix material they produce. David Shine, Paul Harcourt, and Sidman are now in the process of defining the role of the distal stump and of identifying which cellular components and their products are essential to add (to a blind-ended tube lacking a distal stump) to allow fiber regeneration.

Although significant progress has been made in the field of peripheral nerve regeneration, inducing regrowth of central nervous system axons has been especially difficult. Until recently, most researchers believed that differentiated central nerves were incapable of regenerating after an injury. However, Aguayo, M. Vidal Sanz, and their colleagues at McGill University capitalized on the greater regenerative capability of the peripheral nervous system to induce fiber outgrowth in the central nervous system (*4*).

In recent work with Susan Kierstadt and Michael Rasminsky, they removed a section of the optic nerve (a CNS fiber tract that carries visual information from the eye to the brain) and replaced it with a section of sciatic nerve (PNS). New CNS optic nerve fibers grew into the PNS graft. The greatest regenerative response occurred when the CNS lesion and the PNS graft were made very close to the cell bodies of the retinal ganglion neurons, whose axons form the optic nerve.

Aguayo's group reported their latest results at the Society for Neuro-

*The Annual Meeting of the Society for Neuroscience was held in Dallas from 20 to 25 October. The Neurobiology of Disease Workshop on Regeneration preceded the meeting.

science meeting. They showed that severed axons of adult rat retinal neurons can regenerate through an entire 20- to 30-mm PNS graft, a distance about twice the length of a normal optic nerve. Significantly, they were also able to demonstrate some restoration of physiological function.

By using light to stimulate specific receptive fields in the retina, they produced electrical activity in regenerated axons that had been teased from the PNS grafts. The electrical responses were "indistinguishable from normal responses." Although this does not indicate that sight was restored in the experimental animals, these new findings are especially significant. Additionally, it seems to be important to the growing axons that they find their normal target in the brain, because axons denied access to target tissue eventually lose physiological function.

The mechanism by which peripheral nervous tissue encourages CNS fiber outgrowth is still an open question, but differences in the chemical composition of the substrates for the two tissues appear to be crucial. Support for this view comes from Thoenen and Martin Schwab, also in Martinsreid, West Germany. They report that neurons will choose a peripheral nerve substrate rather than a CNS substrate for extending new fibers, possibly because adult central nerve tissue contains certain substances that inhibit growth (*5*).

Schwab and Thoenen used dissociated rat sympathetic or sensory ganglion cells growing in culture and gave the neurons a choice. They could extend axons along a "bridge" of either optic (CNS) or sciatic (PNS) nerve explants. When possible, axons grew along other axons, forming bundles of fibers. Besides this tendency to "fasciculate," the consistent choice for

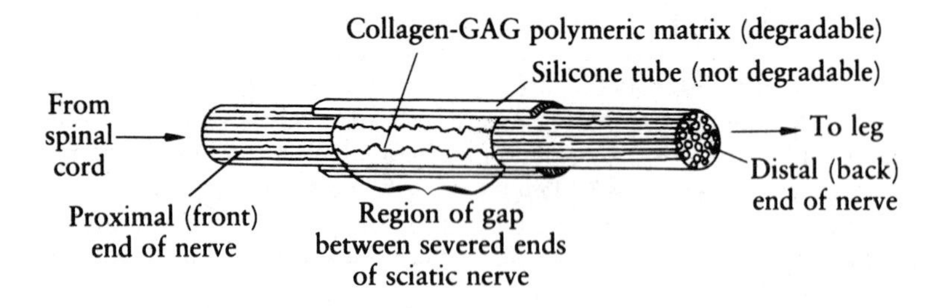

Side view of the eye and optic nerve showing sciatic nerve graft used by Aguayo and his colleagues.

both sympathetic and sensory axons was to extend along the peripheral nerve explant and to avoid the CNS explant completely.

Specifically, nerve fibers grew along Schwann cell surfaces and the basal lamina. They made the same choice even if the explanted tissues were not living. According to Schwab and Thoenen, this points to "the presence of a nonpermissive substrate in the CNS" that persists even after the cells producing it have died.

The tissue culture conditions were optimized for extensive fiber outgrowth by adding nerve growth factor. Because the growth conditions were so ideal, Schwab and Thoenen cite the failure of axons to grow along the CNS explant as evidence that, "in the differentiated CNS, inhibitory substrate molecules should be considered."

In presentations at the Neurobiology of Disease Workshop on Regeneration and also in a symposium at the Society for Neuroscience meeting, Paul Patterson summarized further evidence that certain materials outside cells, in the extracellular matrix, are essential for fiber outgrowth.

Two components of the extracellular matrix that appear to support neuronal outgrowth are the protein, laminin, and a proteoglycan, heparin sulfate. These materials form a complex in the extracellular matrix, which is associated with Schwann cells in the peripheral nervous system but not in the central nervous system. Patterson explained that, in order to grow, axons must be able to both attach to their substrate and detach from it.

Citing work done in collaboration with Randall Pittman, Patterson offered an intriguing hypothesis by which growing axonal tips control their own behavior, providing there is an appropriate substrate.

Pittman studied the elongation of rat sympathetic ganglion axons growing in culture (6) and reported his most recent results at the Society for Neuroscience meeting. Axons appear to modulate their own growth by producing substances that alter their attachment to the substrate. Patterson and Pittman suggest that this modulation is achieved by a balance between certain enzymes and their inhibitors that exist in the local environment of growing neurons. With such a balance, a cycle of attachment and detachment of growing axons could be explained.

Thus, there is increasing evidence for cell-cell interactions and cell-substrate interactions during regeneration of differentiated nerve tissue. The interactions may be positive, or negative, or absent. What was not apparent before is that nerve cells in the mammalian central and peripheral nervous systems can regenerate if they are given the right environment, and that some of their functional characteristics can be preserved. Ultimately, it must be shown that regenerated axons connect in a specific manner with their target tissues, and that normal stimulation of the regenerated nerve produces an appropriate response.

References and Notes

1. G. Lundborg *et al., Exp. Neurol.* **76,** 361 (1982).
2. R. Madison *et al., ibid.* **88,** 767 (1985).
3. C. Da Silva *et al., Brain Res.* **342,** 307 (1985).
4. M. Munz *et al., ibid.* **340,** 115 (1985).
5. M. Schwab and H. Thoenen, *J. Neurosc.* **5,** 2415 (1985).
6. R. Pittman, *Dev. Biol.* **110,** 91 (1985).

Book Review

Reading book reviews is one way that professional scientists keep up with their fields. It's also an excellent method for the student of biology to sample the range of current theory and research. Reviews of biology books appear in serious magazines like the *Atlantic* and in book review journals like the New York *Times Book Review* and the *New York Review of Books*. But they can be found in far greater numbers in scholarly journals like *Science, BioScience*, and *The Quarterly Review of Biology.*

Reading book reviews is one thing, writing them is another matter. Because it exposes the student to the primary literature in the field, the assignment to write a book review is common not only in biology courses but also in psychology, education, history, and many other disciplines. Although the specifics of the assignment may vary, professors usually want to help you develop analytical skills in reading, to give you practice in presenting main ideas and supporting them with a line of reasoning, and to encourage you to make a critical assessment of the book.

A good book review for a biology course should

1. Identify the book completely: author or editor, full title, publisher, and place and date of publication.
2. Describe the subject and scope of the book.
3. Give information about the author, focussing on his or her qualifications for writing this book. (Ask the reference librarian for help in finding this information.)

4. Outline or summarize the thrust or argument of the book, giving the main pieces of evidence to support the author's position.
5. Tell whether, in your judgment, the author satisfactorily supports the thesis or argument.
6. Connect the book to the larger world: by explaining the ramifications of the argument or material, by assessing the value of the book, or by placing it in the context of public issues or of other current books on the subject.
7. Relate the book's subject or thesis to the academic course.

How does an actual book review incorporate these elements? To find out, read the following review of *Trial and Error* from *Science.*

The Curricular Arena[4]

THOMAS JAMES

Educational Studies Program, Wesleyan University, Middletown, Connecticut 06457

Trial and Error. The American Controversy over Creation and Evolution. Edward J. Larson. Oxford University Press, New York, 1985. viii, 222 pp. $17.95.

The rise of modern science and the spread of mass public schooling, though simultaneous, have not always been mutually reinforcing. In the United States, the rules that guide the selection of what is to be transmitted to children through public education are essentially political: their legitimacy comes from state constitutions, statutes, and legal precedent that tie public education explicitly to the authority of the government, which is based upon majority rule and consent of the governed. In contrast, the rules that shape the quest of science are methodological; their legitimacy springs from a set of conventions that experience has demonstrated to be successful for systematically observing, generating hypotheses, devising experiments, and perfecting theories about reality. Though it is true that scientific inquiry can be influenced by political considerations, the scientist is still free to inquire in ways the schoolteacher is not free to teach.

Because of this difference, the United States is a nation in which science education has the possibility of becoming a contradiction in terms. At

[4]From *Science* 230:1266–1267. Dec 1985. Copyright © 1985 by the AAAS. Used with permission.

times, basic modes of inquiry considered essential in science and social science, such as evolutionary thought in biology or theories of human development in psychology, have been inadmissible to the political culture of schooling. At other times, groups have sought legislation to enforce the teaching of preferred values and beliefs under the guise of science. This was the case with "scientific temperance instruction" at the beginning of the century and "creation-science" in recent years.

Beyond these obvious cases lies a larger disjunction between science and education. Schools teach codified knowledge under the wary eye of the public. In the political life of schooling, the continual strife over what should be instilled in children as knowledge tends to obstruct the play of open inquiry, favoring instead a consensus based upon accepted social values. Schools are purveyors not of science but of "public science," observes Edward J. Larson in *Trial and Error: The American Controversy over Creation and Evolution*. Public science, or "publicly supported science teaching and related activities," represents a compromise between scientific thought and popular opinion.

Such a compromise often comes about through legislation and litigation. In "public science," legal restrictions are the ultimate arbiters of disputes over school curricula. This angle of vision is the key to Larson's interpretation of science education in American high schools. The central concern of the book, evolution versus creationism in school curricula, is by no means a new story; numerous books and articles on the subject have appeared over the past decade, and there are no surprises in the evidence marshaled for this study. Nonetheless, *Trial and Error* complements recent studies by showing how legal rules and procedures, along with shifts in popular opinion, helped frame argumentative strategies and shape public decisions over the years. No other study succeeds so well at portraying the development of political argument and legal reasoning in historical context.

Of particular interest are the carefully drawn connections among celebrated cases, social movements, scientific culture, public perceptions, and the growth of schooling. The author brings to the task a fitting blend of training and perspective, possessing both a law degree and a doctorate in the history of science and having served as counsel for the Committee on Education and Labor in the U.S. House of Representatives. The main contribution of the book is that it traces clearly the legal controversies surrounding evolution and creationism in American high schools, but readers will also enjoy a vivid retelling of personal credos, political machinations, pedagogical developments, and other historical circumstances surrounding the vicissitudes of "public science" in the schools.

Most of the controversy has centered on the language of textbooks. Larson points out that one reason for concern was the tremendous expansion of the American high school, whose enrollments approximately doubled every decade between 1890 and 1940. An analysis of multi-edition textbooks between 1859 and 1920 shows the widespread adoption of evolution in the science curriculum during the early years of high school expansion. After 1920, these parallel developments met head on with the growing insecurities of disaffected groups hoping to preserve traditions of evangelical Christianity and strict morals against what they saw as the onslaught of secular and scientific modernity.

Describing the tension of "public science" and popular opinion against this background, Larson brings to life the strategies of the anti-evolution movement of the 1920's as it pressed for legislation in more than 20 states, succeeding in five. In an argument that may claim too much for the anti-evolutionists, he contends that the campaign against science in the schools was a product not only of fundamentalism defined as a "religious movement for biblical literalism" — but of wider currents of progressive reform that were sweeping the country. The historical account draws from newspapers, speeches, tracts, books, the papers of William Jennings Bryan, and an extensive array of secondary literature. Adding to these sources the papers of Clarence Darrow and a complete trial transcript, the narrative goes on to summarize the ideological posturing and national attention that converged in 1925 on the *Scopes* trial in Tennessee, in which that state's law banning the teaching of evolution in the schools was upheld.

Evolution sank out of sight in high school textbooks for several decades after the 1920's. After years of neglect, the issue suddenly surfaced again by way of new science curricula developed with federal funds after the launching of Sputnik I by the Soviet Union in 1957. Perhaps the most useful contributions of *Trial and Error* is its depiction of the sequence of actions that first won legitimacy for evolution in school curricula during the 1960's and then maintained its primacy against creationist challenges through the mid-1980's. Using court records, organizational publications, newspapers, books and articles, statutes, legislative journals, and personal interviews, Larson provides a lively account of recent developments.

The ruling of the U.S. Supreme Court in *Epperson v. Arkansas* (1967) struck down that state's anti-evolution law. Since that time, judges have read creationist challenges to evolutionary teaching as unconstitutional attempts to establish religion in state institutions. Creationists have responded by devising new legal strategies as well as research and organizational base for promotional activities. The scientific establishment, educators, and legal action groups, in turn, have become better organized for explaining the difference between science and non-science in school curricula.

More subtly, according to Larson, judges have responded to popular opinion in finding anti-evolution and creation-science statutes repugnant to "the modern mind." Judges have shown deference toward greater public acceptance of the methods and social meanings of science in the United States. In so doing, it might be added, they have acknowledged a vital connection between scientific inquiry and the civic and social purposes of education in a democratic society. They have protected that connection against groups demanding a similar legitimacy for their own preferred systems of belief. In the political calculus that underlies "public science," the principle of majority rule has shifted the balance of power in controlling school curricula since the 1920's. Several strands of historical change help to explain this shift, notably demographic movements, political realignments, and higher levels of scientific education in the populace. Creationists, for their part, have shown an awareness of the shift as they have attempted to present traditional doctrines in scientific garb and, as a minority, to claim that without "equal time" their rights are being infringed upon, an argument that so far the courts have rejected.

What the author finds most interesting, and describes well, is the resourcefulness of the proponents on both sides as they have countered each other's strategies repeatedly in legislative chambers and courts of law. Since the contention is not likely to cease, this book merits attention for its many insights into the dilemmas of science education in a democratic society.

Not all reviews, however, are this long. Reviews in *Science Books and Films,* for instance, are generally much shorter. They must still provide the reader with enough information to decide whether to read the book.

HYDE, MARGARET O., and LAWRENCE E. HYDE. Cloning and the New Genetics. (Illus.) Hillside, NJ: Enslow, 1984. 128pp. $10.95. 83-20727. ISBN 0-89490-084-6. Glossary; Index; C.I.P.

This book for laypersons attempts to explain the "new biology" with short sentences and a minimum of jargon. It covers aspects of cloning, such as recombinant DNA and molecular cloning, monoclonal antibodies, and applications to medicine and agriculture. Line drawings and photographs (none original) enhance the text, but the electron photomicrographs lack clues to their magnification. The reading list is well chosen and up to date. However, this book cannot be recommended because it contains serious errors of fact: in chapter one, after correctly defining a clone and pointing out that all organisms in a clone contain the same genetic material (p. 13), the authors incorrectly equate parthenogenesis with cloning; they consider the progeny of a clutch of unfertilized fish or reptile eggs to be a clone (p. 17–20)! Since cloning is the main subject of this book, such an error is inexcusable. There are other misconceptions and ambiguities, such as "the armadillo, a mammal, reproduces asexually routinely" (p. 18). There is a misleading description of the manufacture of human growth hormone (p. 62) and faulty definitions of nuclear transplantation and of tissue culture (in the glossary). After a description of embryo transplants in cattle and humans, the authors mention that "people often confuse this technique with cloning" (p. 22). True, and this book is likely to add to their confusion. — *I. J. Lorch, Canisius College, Buffalo, NY.* Concurring remarks provided by *Matthew J. Temple, Nazareth College, Rochester, NY*[5]

WRITING 3.4: READING A BOOK REVIEW. If you haven't already done so, read the review of *Trial and Error* on pages 53–56. As you read, look for the elements of a good book review explained above and mark them in the margins. For instance, mark with the number 2 the paragraph or sentence where the review describes the subject and scope of the book. Are any of the elements omitted? Can you think of reasons for the omission? Which of the elements receive the most extensive treatment? Can you suggest why? Doing these things should give you a better idea of the way one reviewer, at least, fulfilled his obligations to his readers.

WRITING 3.5: WRITING A BOOK REVIEW. Choose a recent book within the scope of this course (your professor may provide a list of titles or direct you to the suggestions for fur-

[5]*From Science Books and Films*, Vol. 21, No. 1, 1985, page 23. Used with permission.

ther reading in your textbook). Before you read the book, review the guidelines given earlier. Knowing what you are looking for will help you read faster and more effectively. When you write your review, refer to the guidelines frequently, but don't follow them slavishly. Let your own responses to the book show through.

[4] *The Term Paper*

PREVIEW: *Practically every student has to write a term paper at one time or another. Follow the process approach explained in this chapter for superior results.*

I am never as clear about any matter as when I have just finished writing about it.

—James Van Allen

"Twenty-five percent of your grade will be based on a term paper which is due the last day of class."

Has this happened to you? No other instructions. It is up to you to pick the topic, research the literature, and write the paper. Are you prepared for this challenge? Although you may think you're not up to it, if you have read Chapters 1 and 2 and have been faithful about keeping your journal, you are far better prepared than you might think.

Perhaps your first reaction to this assignment is frustration —

that wouldn't be surprising. "This is a course in evolution," you mutter, "and I have to find one topic worth 25 points. What do I know about evolution this early in the course? Except for Darwin's *Origin of Species*, which I had to read in high school biology, I have never even thought about it. Oh, I know that some guy named Gould writes essays in *Natural History* (you told us that) and my ecology professor last semester told us to read a book called *Ge* or *Gaia* or something like that. Maybe I could read that now and write about whether I agree with him or not. . . ."

WRITING 4.1: JOURNAL WRITE. Frustration and uncertainty can be productive — providing you channel that energy. If you receive an assignment like that described above, writing about it in your journal can be of value. It's certainly acceptable to begin by complaining. But when you get that out of your system, try to focus your energy on the task at hand: preparing to write a term paper. If you get only as far as the imaginary student above did, you will have made a great deal of progress.

PREWRITING THE TERM PAPER

First of all, just what is a term paper? A better name might be **monograph,** for this is an objective paper based on a literature review. Writing a term paper is a challenge. It is much more than just gathering information, shuffling it into some order, and putting it into your own words. The paper should be focussed — you should have a statement to make. Regardless of whether the choice of topic is yours or the professor provides a topic list, you must have a purpose in writing the paper. Writing to satisfy an assignment is not a good purpose, although it is a reason. Many students write papers for all the wrong purposes: to show that they know the format, that they can compile a long bibliography, that they know how to use quotations. Others, more fortunate or perhaps wiser, write papers to learn more about a subject or to state their own views and argue a point. Wanting to learn something, to persuade the reader to your point of view, or to challenge a theory or concept — these are purposes that make writing a paper an exciting prospect.

Not all term papers are argumentative. Many are writen to in-form by presenting an extensive review of the literature. "Rice," a paper by Swaminathan (1984), is an overview of the role of this grain in the world food market. It explains the history of rice, genetic research to provide a better plant, and future uses of rice to alleviate starvation. This kind of paper, the sort often submitted as student term papers, can be most useful to readers. Or it can be boring to write and even more boring to read. The writer may learn something about the subject, but little about the process of synthesis or the excitement of discovery as he or she tries to prove a point. Two recently submitted student papers exemplify this problem: "A summary of cetological study" and "The basic physiology of the lymphatic system" were long (23 and 62 pages respectively) papers full of information but with little more to offer than a rearrangement of existing knowledge. The reader's reaction is likely to be "So what?" Focussing the latter paper on the lymphatic system's role in cancer or AIDS, for instance, would add greatly to reader interest.

"So what?" isn't the reaction to the student paper, "D.D.T. and avian reproduction: Relationships between the use of chlorinated hydrocarbon pesticides and avian reproduction declines." The author argued for the connection between these pesticides and avian reproduction and a ban on D.D.T. use. A similar purpose motivates a book mentioned earlier, *Gaia: A New Look at Life on Earth* (Lovelock 1979). Lovelock offers an entirely new approach to our understanding of the role of organisms, especially microbial, and the establishment of the earth's atmosphere.

Your purpose is intimately linked to the topic you select, the audience defined, the voice chosen, and what you will learn from this assignment. That, after all, is your professor's purpose in assigning the paper — to provide you with the opportunity to learn, both from the information you gather and from the process of writing about it.

During the prewriting phase you have three tasks to perform: choosing a topic, deciding how to analyze and present the information, and taking a position on the subject.

Choosing the Topic

Start early. Don't wait until the paper is nearly due. Once enough of the course has passed so that you have a grasp of the material, start to select a topic. Allow yourself at least a month to prepare a term paper. More time is better: you can use it to revise what you have written.

If your professor provides a list of possible topics, this part is easy. If, however, you have to find your own, life becomes more difficult. Look to your journal. Have you been keeping a Topics List as part of your daily entries? Can you use any of these topics in this course? If the list is of no help, then you must start fresh.

If you find, or are given, a general topic, too broad to cover in the number of pages and time available, try the technique called **clustering.** Clustering is a simple way of discovering the elements of a subject and the relationships between those elements. Begin by jotting down the main subject, then draw a circle around it. Next, surround the main subject with the central ideas or major divisions. Circle these and draw lines connecting them to the subject. Finally, jot down near each idea or division any details, examples, or further divisions. Circle these too and draw lines connecting them to their "parents." Figure 4.1 illustrates one student's clustering of ideas for a paper on the broad subject of acid precipitation.

From the clustering activity shown above, the student developed a list of possible topics for her term paper:

effects of acid precipitation on human physiology

chemical mechanisms of acid precipitation

sources of acid precipitation

effects of acid precipitation on forest land

impact of acid precipitation on agriculture

WRITING 4.2: CLUSTERING. Develop a cluster like the one shown above. The subject is up to you, but the activity is likely to be more useful if you choose a subject that you are considering for a paper. When you have finished the cluster, try to list four or five possible topics for a ten- to twelve-page term paper.

Previewing the Literature

Before you settle firmly on a paper topic, you should determine if you can find enough material to sustain a paper of the length re-

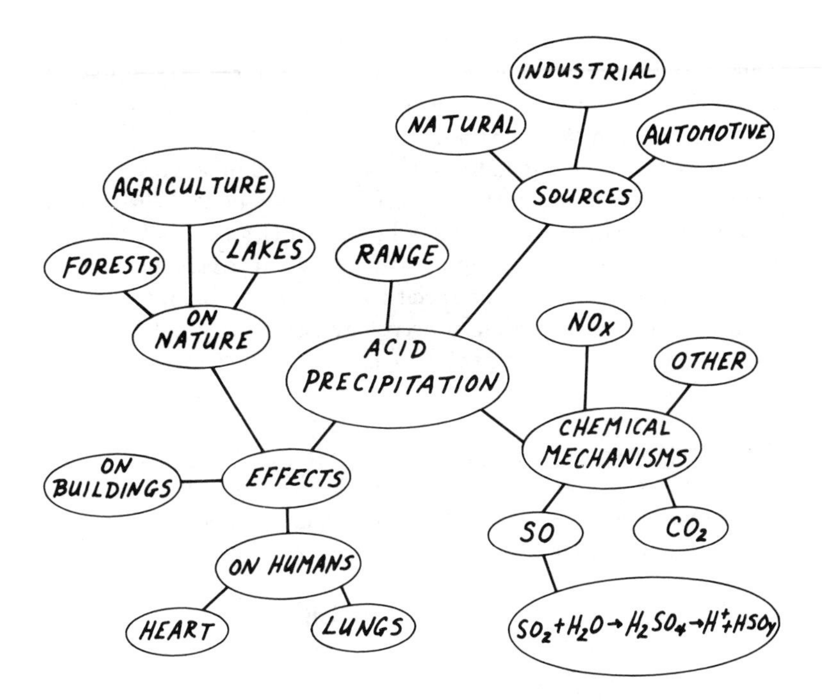

Figure 4.1 *Clustering of ideas for the topic Acid Precipitation.*

quired. You do this by previewing the literature. Begin with several topics that interest you. Be sure they are relevant to the assignment. Then search to learn how extensive the literature is on each of these topics. If you can't find at least eight to ten references for a topic, you will probably have difficulty covering the subject thoroughly. Chapter 7 gives details on building a bibliography. Don't overlook the obvious: references in your text book and in articles you find. References in one source can lead to others. If yours is a large university with an extensive library, finding the reference material will probably not be difficult. On the other hand, if your library sources are limited, it may be harder to find eight or ten references for some topics. In choosing possible topics, then, keep in mind the research interests and courses taught by your professors, who will tend to ask the library to purchase books and reference texts in their areas.

If you restrict your topics to these fields, finding enough material will be easier. Once you have exhausted your library's resources, it is sometimes helpful to ask your professors if they have additional materials relevant to your topic.

As you preview the literature, you are building a working bibliography and skimming the materials for potential. Write down complete references for all the books and articles you find. Using a separate 4″ × 6″ card or slip of paper for each source is common practice. (See Chapter 7 for specific instructions.) Skim these materials as you go, recording on separate cards the kinds of information they contain, but don't take extensive notes just yet.

WRITING 4.3: PREVIEWING THE LITERATURE. Prepare a list of three topics and compile an 8- to 10-item working bibliography for each. Also make a list of additional topics found during your reading. Save these lists in your journal.

WRITING 4.4: SELECTING YOUR TOPIC. When you have completed Writing 4.3, select one topic to research and write about. Make a list of what you know about your topic already. Then make another list of questions you want answered about this topic. After each question, write the source where you think you might find the answer. This writing should help you to see where you stand and what additional research you need to do.

Reviewing the Literature

While you were choosing a topic, you read rapidly through the available sources looking for references. Now it is time to sort your reference list to find the useful items. Take your reference cards and note the relevant ones. Return to the sources and read them for content.

Good note-taking is a lifesaver when writing the body of your paper. You may have already indicated on some cards where you want to use them in the paper. This will help you organize. When making notes from your readings, you should rephrase in your own words everything you put on the cards. If you do want to copy something verbatim, be sure you put it in quotation marks. Then when you use it in your paper, you will know it is a quotation and treat it accordingly. See Chapter 8 on the proper style for citations in the body of your paper. As you do this reading and note-taking, be alert to additional sources you might have missed in your earlier search. Continue to build your working bibliography.

> WRITING 4.5: REVIEWING THE LITERATURE. Follow the steps in the preceding paragraphs. Your end-products will be a set of reading notes from all your sources and a working bibliography on note cards.

Analysing the Topic

Once you have chosen a topic and reviewed the literature, you are ready to begin your analysis of the topic. This is when you select an audience and a voice for your paper and determine your approach to the topic, if you haven't already done so. Who will you write for and how will you present the material? Three possible ways of analyzing a topic are chronological, issue-oriented, and opposing viewpoints.

The **chronological-review** approach allows you to arrange all your information in a time sequence. Adrian Desmond (1977) used this approach in his book *Hot-Blooded Dinosaurs*. A similar technique is found in *The Double Helix* by James Watson (1968). Comparing these two books also allows you to see the different ways in which each author uses voice. Desmond employs a lecture style, while Watson uses a letter-writing style. The potential pitfall of this approach is the "so what?" problem mentioned earlier. A sharply defined purpose will avoid the problem.

If you are using the index card system of note-taking, it is fairly easy to write your first draft. Organize your cards in chronological order and begin to write using their sequence as your outline.

The **issue-oriented** approach is generally used to present a variety of viewpoints or details before arriving at a specific conclusion or synthesis. This is the approach you are most apt to use in writing about current events or discoveries. Examples are *The Naked Ape* by Desmond Morris (1967), *On Aggression* by Konrad Lorenz (1963), Louise Young's *Power Over People* (1973), and Elaine Morgan's *The Descent of Woman* (1972).

To prepare a first draft in this style from your note cards, organize the cards in what you believe is an understandable sequence. Make an outline or table of contents. Then write the draft and review it for clarity. If it doesn't work, rearrange the notes and try again.

The Biological Clock: Two Views by Brown, Hastings, and Palmer (1970) is a good example of the **opposing viewpoints** approach to preparing a monograph. Present each view, along with supporting materials and counterarguments. It is then up to you to interject your own conclusions and support them. When using this style it is helpful, as you take notes from the literature, to write down your immediate reactions to the article you are reading (e.g. "this doesn't seem to agree with the research of . . . ," "perhaps I can use this to support . . . ideas").

Presenting your own views is important in most term papers. Do not become simply a conduit of second-hand information. Try to select a topic which allows you to make decisions and take positions. This is not always possible, or even desirable, but is a good rule to keep in mind when choosing between two topics. If you can become involved with the subject, it will be more interesting to research and write — and more interesting to read.

Paper Styles

In writing your paper you might use one of several styles. Stephen Jay Gould's essays in *Natural History* magazine are similar to English essays. The voice is personal and informal. When reference is made to sources, the complete reference appears in the body of the essay. Articles in *Scientific American* are also in an essay style but each issue contains a one-page bibliography section listing a few references for each article. If you were to write for *American Scientist*, you would follow the CBE guidelines (see Chapter 8). This is the format most of your term papers will follow. Check with your professor for any specific requirements.

WRITING 4.6: ORGANIZING YOUR MATERIALS. Arrange your ideas and materials in a tentative outline or table of contents. This will also require you to decide on the approach you will take to the paper: chronological, issue-oriented, or opposing viewpoints.

THE FORMAT OF THE TERM PAPER

Most term papers consist of the title, the body of the paper, and the list of references cited. Sometimes the body is preceded by an abstract.

Title

The title of a term paper need not be as specific as that of a research report but a descriptive title does help readers decide if the paper is something they are interested in reading. Ordinarily you would have a separate title page (Figure 4.2) although your professor may ask you to place the title at the top of the first page of text.

Auth's title, while descriptive and in proper format, could be improved: "Morphology and Physiology of the Golgi Apparatus" or "Structure and Function of the Golgi Apparatus" would have been better. Your working title may be rather general. After the paper has been drafted, you should rewrite the title to fit the true subject of the paper.

The Abstract

The abstract for a term paper need not be as long or thorough as that written for a research report. Generally a sentence or two outlining the paper is enough. Both *Scientific American* and *American Scientist* magazines contain brief, well-written abstracts; look at a few of them before you write your own. The abstract is usually written last, after you have determined what your paper is truly about. Here is the abstract for the Golgi apparatus paper:

    Until the advent of the electron microscope,
the appearance of the Golgi apparatus was dismissed
as being a mere artifact, the result of harsh

GOLGI APPARATUS:

MORPHOLOGY AND FUNCTION

Presented for

Senior Evaluation

to

Biology Department

St. Michael's College

by

John T. Auth

1971

Figure 4.2 *Title page.*

staining techniques. Investigations with the electron microscope have shown the Golgi apparatus to be a separate, distinct organelle of the cell cytoplasm. Many probable functions have been attributed to the organelle but research thus far has shown

```
the organelle to be the place where glycoproteins
are formed and packaged.
```

The Body

The body of your term paper will generally have three parts: an introduction, the main section, and a summary. The introduction is used to tell the reader the scope of your paper and what you intend to write about. The main section presents your case, using one of the formats for analysis discussed earlier. The last few paragraphs summarize the paper and present your views and conclusions with support. Near the end of this chapter you will find sample pages showing parts of Auth's paper.

Tables and Figures

Tables are lists of numbers and **figures** are diagrams or illustrations. Each is a different way of summarizing information relevant to your presentation. Every table and figure must have a title. Titles of tables appear above the tables. Titles of figures are placed below. Tables and figures are numbered separately and consecutively as they appear in the paper.

If, in preparing your paper, you use copies of tables and figures from some of your sources, it is necessary to credit the original authors. Do this by citing the author and date of the source in parentheses after the title of the table or figure (e.g. "From Leary 1957"). The following example is taken from Auth (1971). If you make any changes in the table or figure, use the phrase "After . . ." rather than "From. . . ." For additional help in constructing and using tables and figures, see Chapter 6.

The List of References

While writing your paper, you refer to various articles and books to support the views being discussed. The references section of your paper is where you will list in alphabetical order all the references mentioned. If you have complete note cards, this section is easy to write. Chapter 8 describes how to prepare this section and compares it to a bibliography. Some professors require bibliographies, so check with your professor.

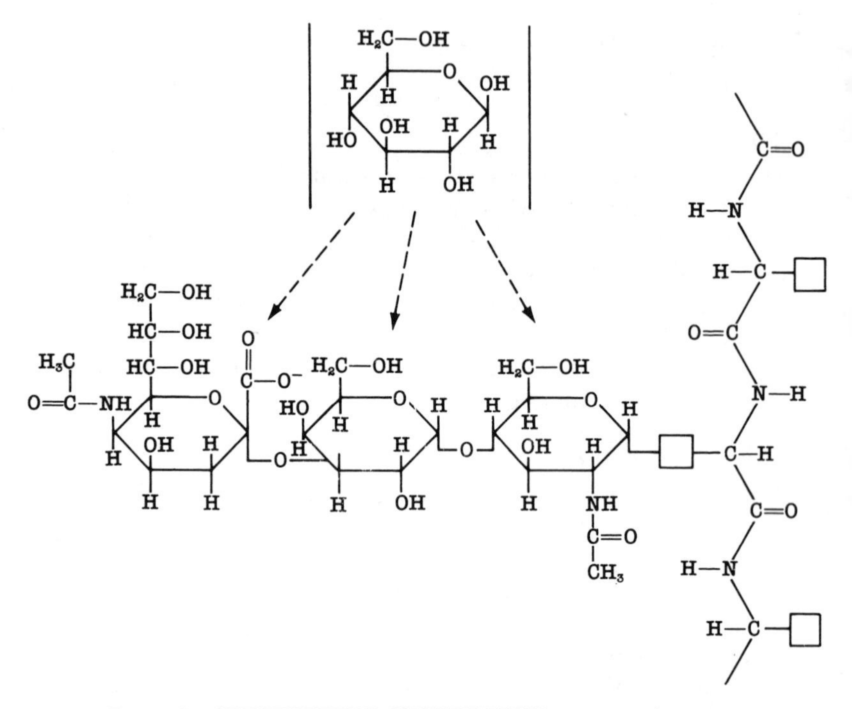

Figure 1. **HYPOTHETICAL GLYCOPROTEIN** consists of a protein backbone with amino acid side chains to which carbohydrate chains, one of which is shown here, are attached. The composition of the sugar chains varies; here the units are N-acetylglucosamine (attached to protein), galactose (middle) and sialic acid (left). These simple sugars can be obtained by modification of glucose (top) supplied to the cell from the blood. (From The Golgi Body, Neutra, M. and C.P. LeBlond. Copyright © 1968 by Scientific American, Inc. All rights reserved.

Figure 4.3 *Example of a student's use of a borrowed figure and the proper citation.*

Sample Pages of a Term Paper

Four sample pages from the student paper on the Golgi apparatus are reproduced on the following pages. Each page shows the begin-

INTRODUCTION

In 1898, the Italian microscopist Camillo Golgi described a network in the cytoplasm of Purkinje cells of the brain of the barn owl treated with silver salt which he called the internal reticular apparatus (Brown and Bertke, 1969). These previously unknown bodies were extremely small structures that appeared to be composed of minute plates and threads. The eminent Spanish histologist Santiago Ramón y Cajal had seen the same curious bodies in silver-stained cells several years earlier, but he had not reported the observation at the time because as he wrote later in his memoirs, ''the confounded reaction never appeared again!'' The mysterious structure was christened the Golgi apparatus (Neutra and Leblond, 1969).

For their studies of nerve cells (which were conducted independently) Golgi and Ramón y Cajal were jointly awarded the Nobel prize for physiology and medicine in 1906. Although the Golgi apparatus was given much attention, for many years there were doubts about its existence, some investigators dismissing the reported observation as a mere artifact.

Figure 4.4 *Student paper — Introduction.*

MORPHOLOGY OF THE GOLGI APPARATUS

The Golgi complex, also known by the term dic-tyosome, is a system of unit membranes found in a wide variety of plant and animal cells. Its mem-branes lack attached ribosomes and it may show con-tinuity with the endoplasmic reticulum but it is more often discontinuous. The organelle may be viewed with a phase-contrast microscope after it has been stained with osmium or silver containing dyes (Swanson, 1969).

So far as light microscope studies are con-cerned, the complex in animal cells can at least be differentiated into two parts: (1) an outer osmio-philic component, and (2) an inner osmiophobic com-ponent. Electron microscopy studies have indicated that the complex has a generalized structural pat-tern made up of a system of membrane-bound vesicles of varying size, associated with smooth membrane elements arranged in more or less parallel fashion. Present evidence suggests that this complex is a separate organelle of the cytoplasm and not a part of the endoplasmic reticulum. In support of this view, the following facts may be cited: (1) the. . . .

Figure 4.5 *Student paper — Morphology.*

FUNCTION OF THE GOLGI APPARATUS

Until very recently, not much was known concerning the functioning of the Golgi apparatus. Most of the early work with the organelle dealt with descriptive morphology. However, in the past ten years or so, much investigation has been done dealing with Golgi function. I will list many of its probable functions and describe in detail its role in lipid metabolism.

Because the complex is somewhat difficult to isolate from other cellular components, its exact function is not precisely known. The fact that the complex is most readily revealed in cells which have a specialized function in connection with secretion, combined with the observation that there is a general parallel between the morphological changes in the complex and the level of secretory activity, has led to the notion that the Golgi functions as a region for the collection of specialized products. The association between the endoplasmic reticulum and the complex provides a logical means of transport to the cell exterior (Wilson and Morrison, 1961).

Figure 4.6 *Student paper — Function.*

REFERENCES CITED

Brown, W.V.; E.M. Bertke. 1969. Textbook of Cytology. C.V. Mosby Co., St. Louis.

Burgos, M.H.; D.W. Fawcett. 1955. Studies on the fine structure of the mammalian testis. I. Differentiation of the Spermatids in the cat. J. Biophys. Biochem. Cytol. 1:287–299.

Cunningham, W.P.; Morre, D.J.; H.H. Mollenhauer. 1966. Structure of isolated plant Golgi apparatus revealed by negative staining. J. Cell Biol. 28:169–179.

Essner, E.; A.B. Novikoff. 1962. Cytological studies on two functional hepatomas. Interrelations of the endoplasmic reticulum, Golgi apparatus, and lysosomes. J. Cell Biol. 15:289–312.

Grove, S.N.: Bracker, C.E.; D.J. Morre. 1968. Cytomembrane differentiation in the endoplasmic reticulum–Golgi apparatus–vesicle complex. Science 161:171–173.

Mollenhauer, H.H. 1964. An Intercisternal Structure in the Golgi apparatus. J. Cell Biol. 24:504–511.

Mollenhauer, H.H.; D.J. Morre. 1966. Tubular connections between dictyosomes and forming secretory vesicles in plant Golgi apparatus. J. Cell Biol. 29:373–376.

Montagna, W. 1950. Perinuclear sudanophil bodies in mammalian epidermis. Quart. J. Micr. Sci. 91:205–208.

Neutra, M.; C.P. Leblond. 1969. The Golgi Apparatus. Sci. Am. 220:100–107.

Figure 4.7 *Student paper — References.*

ning of one of the main sections of the paper: introduction, morphology, function, and references. The title page and abstract were shown earlier in this chapter.

WRITING 4.7: DRAFTING YOUR TERM PAPER. Using the outline or table of contents you prepared for Writing 4.6, write a good draft of the body of your term paper. A good draft will not likely be the first one, perhaps not the second either. Work on your drafting until you have the major elements under control: purpose, audience, voice, approach, organization, etc. Don't worry about smaller matters like spelling, punctuation, word choice right now. There will be time for those later.

REVISING YOUR WORK

If you are like many of us, you drafted your paper in parts and under the pressure of time. Only when you have it all down on paper and assembled can you see the whole of your work. That is the time to reread and reflect on what you've written. Our word **revise** means **to see again,** and seeing again is crucial at this point.

But how do you go about this process? One way begins by giving it a rest. If you can put the paper away for a couple of days, you can come back to it with greater objectivity. But whether you can afford the time for that or not, start your revision by rereading the paper from the beginning and finish in one work session. This sustained reading will permit you to perceive the report as a whole, not as the separate parts you may have written over a period of time and then pasted together. When you read this way, you can sense the unity or lack of it.

It is generally agreed that revising should begin with the larger elements or broad qualities, then gradually move to the smaller. No sense trying to find just the right word when in the next step you eliminate that entire sentence. The overall organization of the paper is prescribed: title page, abstract, body, and references cited. Be certain you have all of these parts in the right sequence. Then turn to a careful review of the organization of each section. Consider the logic of the organization: it may be order of time, order of importance, or order of complexity. Does each paragraph logically follow from the one before and lead to the one after? Next look at the sentences within each paragraph. They too should have this relation to one another. Finally examine the words. Eliminate those that are unnecessary. (Would it have been better to write "Eliminate unnecessary words"?) Of course, you will want to check and recheck numbers, tables, and the spelling of technical and scientific terms. Carelessness here can damage or destroy your credibility with the reader.

Many of us who write for publication always ask a trusted colleague to read our manuscripts. That person represents the critical reader who may become confused by our presentation or who may argue with our findings. Better to get the criticism when you can do something about it instead of after the paper is in print or

Checklist for the Term Paper

1. Descriptive title
   Is the title descriptive — that is, does it accurately reflect the true subject and your approach to it?
   Does the title page follow the format shown on page 68 or the format your professor or editor requested?
2. Abstract
   Does the abstract accurately summarize the scope and conclusions of the paper?
   Is the abstract 250 words or less?
3. Body
   Does the body begin with an introduction that describes the purpose and goal of the paper?
   Is the central idea prominently and clearly stated?
   Does the main section adhere to the central idea throughout?
   Is all the important literature included?
   Are all figures and tables correctly titled and numbered?
   Are all sources properly cited?
   Are the concluding remarks or summary statements justified by the literature and the line of reasoning of the main section?
4. List of References
   Are all references cited in the text listed here?
   Are all references in the correct format?
   Is the list in alphabetical order?

in your professor's hands. You can use this system too. Pair up with a friend whose judgment you trust and swap papers. Write questions, comments, and suggestions for revision. Discuss each other's papers. Then revise once more.

Whether you type the final draft yourself or have it done by a professional, proofreading is essential. Do it when you are alert, not at five in the morning after typing all night. Then have the fussiest person you know proof it again. There is absolutely no excuse for submitting to journal editor or professor any work that has not been scrupulously revised and proofread. Your term paper or monograph represents weeks of your hard work. Respect that effort and others will too.

WRITING 4.8: REVISING YOUR TERM PAPER. Follow the advice given above in revising your paper.

[5] *The Research Proposal*

PREVIEW: *This chapter begins with a brief overview of the scientific method. Then you will move through the steps of writing a research proposal from choosing a topic to revising your proposal for submission.*

Have you ever noticed that when you cut an apple or a potato and allow it to stand, the cut surface turns a brownish white? Yet when you cut an orange or a lemon, no discoloration develops. Have you wondered why? Have you tried to find the answer?

Albert Szent-Gyorgyi wondered, then sought and found an answer. His studies of oxidation reactions eventually led to the isolation of Vitamin C, for which he was awarded the Nobel Prize in Medicine in 1937.

THE SCIENTIFIC METHOD

Those first two steps, **making an observation** and **asking a question** based on that observation, are the start of the scientific method. The third step is **formulating an explanation** to account for the differing reactions of apples and potatoes, on the one hand, and oranges and lemons, on the other. This explanation is your **hypothesis.** In our example it might have been "There is a chemical difference between the groups which provides protection against the browning phenomenon."

Now that you have a possible explanation, what do you do with it? You test it by **experimentation.** You will need to design an experiment that will lead you to accept your hypothesis or reject it. If the results of the experiment cause you to reject the hypothesis, all is not lost — you've learned something. If the outcome leads to accepting the hypothesis, and if repeated experiments continue to support it, you have reached the final phase of the scientific method, phrasing your conclusions as a **scientific theory.** The theory states that under prescribed circumstances the observed phenomenon will always be expected to occur.

Steps of the Scientific Method

1. Observing
2. Questioning
3. Stating a hypothesis
4. Designing an experiment to test the hypothesis
5. Formulating a theory

All of this seems fairly simple and straightforward. Why then are we not all candidates for a Nobel Prize? The problem often starts at the very first steps — the observing and the questioning. Many of us make hundreds of observations every day. Some we question, others we ignore. Even trained scientists sometimes have problems with their observations, as the following case illustrates.

The Mystery of the Contaminated Cultures

Let's say you are a scientist trying to find a chemical that will destroy bacteria when placed in the media the bacteria are growing upon. You arrange hundreds of petri dishes with media and bacteria. To various sub-batches of the dishes you apply different chemicals. To other sub-batches you add nothing — these are your controls. All the dishes are placed in incubators and allowed to remain while the bacteria grow into large colonies. You hope that some of the chemicals will interfere with the bacteria's growth, perhaps in fact kill the bacteria.

After a prescribed period of time you open the incubator, remove the petri dishes, and record the results. In most of the dishes the bacteria are growing quite well. The chemicals do not appear to have had any effect on them. Some of the dishes, however, have become contaminated by airborne mold, which has settled on the media and grown. All the bacteria colonies near the mold have been killed. Your experiment has been ruined by your careless handling of the dishes, allowing them to become contaminated. What can you learn from the experiment? What would you do next? Use your journal for your reflections.

The mystery of the contaminated cultures actually recurred often during the early years of this century. Microbiologists were looking for a miracle drug to combat infections. Time after time they found their cultures contaminated by this mold. Time after time they discarded the cultures and started over. Yet their observations of the contaminated cultures didn't trigger the question, why were the bacteria in the region of the molds dying? Only one researcher, an insightful Briton, stopped to ask why. Why were the bacteria in the vicinity of the mold dying? Apparently the mold prevented the bacteria from growing — what did that signify? This researcher was able to perceive the problem differently, as others before him had not. His questions and subsequent experiments led Alexander Fleming to the discovery of the effects of penicillin.

The Dance of the Bees

Supporting or refuting a hypothesis often leads to additional questions, problems, experiments, and discoveries. An excellent example of this aspect of the scientific method is found in the controversy over the language of the bees. Do bees "dance"? Is the dance a language? If it isn't a language, how do the bees communicate?

Karl von Frisch's experiments throughout the 1920s, '30s, and '40s demonstrated that once a scout bee finds a good nectar source she returns to the hive and "tells" her hive mates. According to von

Frisch the information needed for them to find this source of food is conveyed in a dance performed by the returning bee. The dance of the bee tells symbolically what kind of flower to look for, how rich the nectar source, how far and what direction to fly (von Frisch 1967).

Karl von Frisch's experiments and his theory withstood testing for two decades. Then in the 1960s Adrian Wenner (1967) re-examined the basic concepts of the theory. Believing von Frisch's conclusions were in error, Wenner reviewed von Frisch's results, formulated his own question, and proceeded to test its accuracy. Wenner's experiments produced a new explanation of bee foraging and information transmission. Wenner concluded that the bees do not use a dance but simply display a conditioned response to the odor of the flower which has been brought to the hive by the foraging bees. In his view von Frisch was wrong. Adherents of von Frisch's theory, however, refused to accept this new theory.

Within a few years James Gould (1975) reviewed both sets of results and conducted his own research. Through a series of wonderfully inventive experiments, Gould demonstrated that, depending on the specific conditions they were laboring under, the bees might use a dance for information transferal or might respond to the conditioning effects of a particular odor build-up in the hive. Apparently von Frisch had observed his bees during the early stages of foraging — before an odor could have built up in the hives. Wenner made his observations after the bees had had the opportunity to become conditioned to a particular odor. Gould's observations and experimental designs clarified the apparent contradictions between the earlier theories. Reflecting on the heated controversy, Gould cautioned all researchers: ". . . it is difficult to avoid the unprofitable extremes of blinding skepticism and crippling romanticism . . ." when discussing some of the concepts in animal behavior and the fervor with which some scientists hold their ideas.

THE RESEARCH PROPOSAL

Fleming, von Frisch, Wenner, and Gould — each of these researchers had to design an experiment. Then they had to find support for their research. Professional researchers must usually look to private foundations or the government (National Science Foundation, National Institute of Health) for their support. For graduate students, that support may come from an off-campus source or from their departments. Undergraduates ordinarily seek support from their professors. The vehicle used in all cases is some variation of the **research proposal.** This document relates what you plan to do, what you already know about the problem, why you want to solve this problem, how you will do it, and what you expect to find.

CHOOSING A TOPIC

Before you can write a research proposal, you will have to find a topic. Most likely this will be for a class project. Your professor may distribute a list of topics from which to choose. That would make the first step relatively easy. But what if the choice is entirely up to you? In Chapter 2 you learned how to use a journal to compile a list of possible future research topics. If you have been noting ideas for research in your journal, you've got a good start. Get it out and look through it. Skim through the entries, noting ideas as you encounter them. Write them down on a separate topic list. Are any of these topics still interesting to you and appropriate to the assignment? (If none can be used, you might ask where you went wrong in using your journal in the first place. Make a note of your answers and modify your journal-keeping for future use.)

A second source of ideas for topics is your reading. Ideally this should be reflected in your journal too. Otherwise, you'll have to conduct a literature review of an area of general interest (such as animal behavior, water pollution, photosynthesis). As you read, make notes in your journal and compile topic lists.

Still a third way to find a topic is to consult your course textbook. The authors often put suggestions for more research at the end of each chapter, along with a list of references for further reading. Both should be valuable aids to finding a topic.

These are only suggestions; it is up to you to keep an open mind and sharp senses as you seek a topic.

WRITING 5.1: CHOOSING A TOPIC. As the first step toward your research proposal, make a list of topics appropriate to the assigned area for your paper or research. Then take your list to the library and do a literature search of the likeliest topics (see Chapter 7 for suggestions). The search is the start of your working bibliography. The greatest topic in the world is no good if you can't find supporting literature. Discard any topic if you fail to find at least ten references easily. And don't forget to keep your mind open for other topics as you search for references. Often the best papers result from discoveries made while looking for something else. Even if you can't use a newly discovered topic now, add it to the topic list in your journal along with a note as to where you found the information.

When you've discovered a potential topic, apply this test:

1. Is it interesting to you?
2. Does it fit the scope of the course and the assignment?
3. Can you find at least ten library references?
4. Do you understand the material?

WRITING 5.2: PREWRITING THE PROPOSAL. Your aim in Writing 5.1 was to come up with a suitable topic and working bibliography for your paper or research proposal. Now, before you actually write the proposal, respond in your journal to these considerations:

Have you made any observations? (Even observations from your reading are acceptable.) What is your question? Do you have a hypothesis? Write freely about the topic you have selected. What problems does it present that you could examine? What does the literature say about these problems? What statement can you make about the problem? How can you test your ideas? What equipment will you need? Can you do the research and write the results in the time available to you?

Writing in your journal should help you to clarify your thinking about your topic and prepare you to write the research proposal.

THE PARTS OF THE RESEARCH PROPOSAL

A research proposal is basically a detailed explanation of what you propose to do, along with why and how and the projected outcome. Its sections are **working title, introduction, methods and materials,**

statement of impact, and **references cited.** The remainder of this chapter will take you through the writing of these sections, with examples of effective writing and warnings about pitfalls. If you do the writings as you go, you will have a complete research proposal in hand at the end of the chapter.

Basic Parts of the Research Proposal

1. Working Title
2. Introduction
3. Methods and Materials
4. Impact Statement
5. References Cited

Working Title

Fuzzy topic — that's the curse of many a beginning researcher. A topic that is too broad leads the scientist into spending too much time on side issues, neglecting the main line of inquiry. You can avoid this problem by designing a good working title: a brief, but complete description of your proposed research. "Acid Rain," for instance, is a poor working title. It is too broad to give direction to the writer or information to the reader. Much better is "Seasonal variations in precipitation pH." Even this title could be improved by giving the locality and limiting the time frame of the study.

The title, "Biological clocks in crayfish," might sound interesting as the title of a review paper, but it is useless both as working title and as the final title of a research report. It is too broad: it doesn't specify whether this is a behavioral, physiological, or biochemical study. Even the species name of the crayfish studied is lacking. An example of a good title is "A Study of competition between *Tribolium castaneum* and *Tribolium confusum* under selected temperature and humidity conditions." Even this title could be improved by specifying the temperature and humidity conditions.

An effective working title will actually help you focus your thinking and control the direction of your research. But don't forget that "working" means that your title is subject to change. As your research progresses, you may well tighten or refine your title. When you have finished, your working title will have evolved into the actual title of your research report.

> WRITING 5.3: YOUR WORKING TITLE. Write it! Then rewrite it until you have a title that accurately and completely describes your proposed research.

Introduction

The introduction explains what you intend to do and summarizes the relevant literature to support your intentions. If you have taken good notes while reading and have carefully thought through your topic, writing this section is relatively easy. Your organization can be chronological, argumentative, or topical. State your premise clearly and present the information needed to defend your proposal. And be sure to document your references in the style explained in Chapter 8.

The actual length and development of this introduction will depend on how much detail your professor or the funding agency requires. Some introductions are as short as one page, with only a brief summary of the current state of knowledge of the problem. Others, much longer, present an extensive review of the literature. Consult with your professor on this matter.

The following introduction is taken from a student's proposal to the National Science Foundation requesting money to support a summer of research as part of NSF's Student-Originated-Studies Program. Although the introduction is just one page, the proposal was effective: Masse and her team received over $15,000 for their study.

Shelburne Bay, on Lake Champlain, adjacent to South Burlington and Shelburne, Vermont, has been cited by the New England River Basin Commission as one of the more seriously polluted bays on the Lake (NERB 1976) and a region which requires more study.

This proposal is for a study to determine the cause and extent of this pollution and the relation of the bay's water quality to the people living around it. Our approach will be a thorough examination of standard physical, chemical, and biological parameters that relate to water quality and the state of the bay.

There are several sources of effluents into Shelburne Bay. Two municipal secondary sewage treatment plants are the major point sources. The bay also indirectly receives the discharges from two additional municipal wastewater treatment facilities located on its major tributary, the La Platte River. The La Platte is the major non-point source of pollutants, but farms on the western side of the bay are also presumed to be significant contributors of N runoff. Nutrient loading of N and P is due largely to these non-point sources. We hope to identify areas contributing to nutrient runoff and to quantify such non-point source loadings.

Many of Lake Champlain's embayments are rapidly becoming eutrophic as a result of increased urbanization of the area -- this is the case with Shelburne Bay. The New England Basin Study reports that algae blooms are appearing earlier in the summer and are becoming more frequent than in the past. We hope to inquire as to whether this is the naturally progressing eutrophication process or an accelerated one due to increased urban activity in the bay's proximity and the increased agriculture along the La Platte River. (Masse 1980)

This introduction, although short and lacking in details of previous studies, presents concisely the important points Masse needed to make. There was a previous recommendation for study of the bay. The bay has several sources of pollution. Eutrophication (enrichment) is a problem in certain sections of the bay. And, finally, she states what her study group hopes to accomplish — determine a possible link between farming practices and eutrophication.

WRITING 5.4: YOUR INTRODUCTION. After reviewing the guidelines at the beginning of this section, draft an introduction for your proposal. Length should be determined by your purpose and your professor's instructions. For best results, ask a classmate to read your draft and make comments. Then revise accordingly.

Methods and Materials

In the introduction you explained what you intend to do. The methods and materials section provides specific details on how (i.e., when, where, and under what conditions) you will conduct your experiment. Summarize the experiment in a narrative, not a list. If you are tempted to list all equipment (e.g., 24 beakers, 2 pipettes, etc.), don't. However, if plants or animals will be used, tell how many are required, how you will subdivide them, what tests will be performed, and what special conditions are called for. Although you may not have worked out all the details, be as precise as possible. Anticipate the reader's questions; anyone should be able to duplicate your experiment by following your directions. You need not, however, give full treatment to tests or procedures that are routinely found in a handbook or are universally understood. The opening paragraph of Masse's methods and materials section illustrates this.

Our techniques will include biological, physical, and chemical water quality tests following

```
procedures found in Standard Methods (APHA 1975).
Samples will be taken at specific stations. There
will be approximately 10-20 of these sites, dis-
tributed along the shore and the middle of the bay.
Specific locations will be determined early in the
summer, based on preliminary testing and informa-
tion from a previous study (Little 1977). We will
choose those locations that provide the best over-
all picture of the bay's character.
```

Later in this section Masse details some of the specific physical and chemical tests to be performed and the reasons for choosing them.

One of the last items in your materials and methods section is usually a statement explaining how you will test your results statistically. Many, but by no means all, experiments require some form of statistical analysis to verify the results. This is especially true of nutritional, genetic, and any other studies where data can be collected for mathematical verification. Commonly used tests include Student's *t,* Chi-square, analysis of variance, and correlation studies. It is always a good idea to discuss your experimental design with a statistician before you begin your study. This can be a friend studying statistics or your professor, some knowledgeable person who can help you decide on the most appropriate test and the best way to record your data so that it can be easily manipulated. Failure to do this may result in much work, lots of information, and no way to tell if it is meaningful.

WRITING 5.5: METHODS AND MATERIALS. Draft the opening paragraph of your methods and materials section, then ask several classmates to read it. Do they understand what you intend to do? If not, why not? Use their comments to improve your directions to the reader. If you still have problems, go to the literature. Read the methods and materials sections of papers in such journals as *The Quarterly Review of Biology,* the *Journal of Limnology and Oceanography,* or those you are reading for your course.

Impact Statement

What will you do with the information you gather from your study? In other words, why conduct this experiment at all? Although you can never be absolutely certain what your results will be, you can project the potential impact of the study. The impact statement explains the importance of this study and tells how its information might be used. The last paragraph of Masse's proposal exemplifies this.

UTILIZATION OF STUDY RESULTS

The final report will be submitted to the Vermont Academy of Arts and Sciences for consideration as a presentation at the Annual Spring Student Symposium. We also anticipate that the towns of South Burlington and Shelburne will not only want and receive a copy of the final report, but that public meetings will be held to present the results. The findings of previous projects have been of great interest to both townships so we expect a similar response toward this study. Mr. Arthur Hogan, Chittenden County Regional Planning Commission, is already very interested in learning what our results show. These results should also prove useful in clarifying a current issue over the rights and appropriateness of a town boat-mooring ordinance in the town of Shelburne.

WRITING 5.6: IMPACT OF RESULTS. Try writing an impact statement for your study. Does it help solve some of the questions that you posed in your introduction? How might a positive answer to your hypothesis improve our knowledge of the problem? How might you explain a negative result? Obviously you can't answer all of these questions without your results in hand, but asking them now forces you to keep an open mind and to test your original premise continually.

WRITING 5.7: REVISING YOUR RESEARCH PROPOSAL. If you have followed the text and the writing assignments, you have a good draft of your research proposal. Because it was written in parts over a period of several days, the parts might not flow together smoothly. Furthermore, your vision of the project might have altered as you proceeded. Now is the time, before final typing, to go through the proposal for coherence to the main plan. When revising any piece of writing, you can save time by starting with the larger elements such as sections and paragraphs. That is, don't worry about small matters of spelling and punctuation until you have the whole piece in good order.

Two other tips you might find useful: one, set your proposal aside for several days before you turn it in. Right now it is your pride and joy; you can't see anything wrong with it. Or, conversely, you might be so sick of it that you can't "see" it at all. Given a short rest, you should be able to view it with fresh eyes. The second pointer you've read in these pages before: ask someone else to read your proposal and comment. Most of us who write all the time do that and treasure a patient, critical reader.

References Cited

The final section of the research proposal is references cited. Here you will list all the studies to which you refer in your proposal. See Chapter 8 for the details of preparing this section.

Some Final Words

In discussing your research proposal we have not touched on one very important item, the budget. Normally in a student project a detailed budget is not necessary. If you submit a proposal to an agency such as the National Science Foundation, you would follow the format they provide.

After your proposal has been accepted, after you have conducted the research, collected the data, labeled the last specimen, and put away your calculator, you have one last task. You have to prepare a final report and submit it to your professor or the granting agency. Writing that report is the subject of the next chapter.

Checklist for the Research Proposal

1. Working title
   Does your title limit the scope of your proposal and explain its purpose?
2. Introduction
   Have you explained the purpose of your study and cited all pertinent sources?
3. Methods and materials
   Have you included all the information needed for another researcher to duplicate your study: where, when, under what conditions? with what organisms?
4. Impact of results
   Do you explain why this study is important?
   Do you explain how the results might be used to answer a question or solve a biological dilemma?
5. References cited
   Are all in-text citations correct?
   Are they all listed in the references section?
   Are all the references accurate?
6. Budget
   If a budget or equipment list is required, is it complete and accurate?
7. Revision
   Do the parts fit together into a coherent whole?
   Have you proofread for mechanical errors?

[6] *The Research Report*

PREVIEW: *Present the results of your research in a well-designed and documented report. In this chapter you will learn how to write the essential parts of the research report.*

> Among scientists are collectors, classifiers, and compulsive tidiers-up; many are detectives by temperament and many are explorers; some are artists and others artisans. There are poet-scientists and philosopher-scientists and even a few mystics.
>
> —Sir Peter Brian Medawar

Have you ever heard someone ask, "Which is more important — what I have to say or how I say it?" The question often implies that the content is more important than the way the ideas are expressed. Not true. In recent months you have read many research reports. Very likely the well-written papers made a much better impression on you than those that were poorly done, even though the contents might have been of equal value. Perhaps you were even tempted to ignore those reports that were harder to read or were presented in an unusual and difficult format.

Accomplished research scientists send their reports to journal editors to be considered for publication. The editor asks other scientists to review the work reflected in the report. The chances of a study actually being published depend on two criteria: the quality of the research and the quality of the report itself. Many competent research studies never reach publication or are at least delayed by revisions because of poor writing and organization. Well-written, easy-to-read, clearly organized reports are the goal of every editor, who is responsible for the quality of his or her journal. Readers faced with dozens of sources of information frequently skip poorly written articles.

The point? If you want to impress readers with the quality of your research, you must first impress them with the quality of your writing.

CHARACTERISTICS OF THE RESEARCH REPORT

You have just spent a month, a semester, or even a year conducting experiments and collecting data as part of your research. Now you have an important statement to make to the world of science (or at least to your professor). How do you go about it?

The normal manner of reporting research results is the **research report.** Because it follows a fairly standardized format, the writer knows what is required and the reader what to expect. As you read scientific papers, do you ever stop to reflect on the style in which those papers are written and the format they follow?

The points that you probably notice are that (1) all the papers have **descriptive titles;** (2) most, if not all, have **abstracts;** (3) a major part of each paper is its **introduction;** (4) each research report has either a **methods** or a **methods and materials** section; (5) all the data are presented in a **results** section; and (6) explained in the **discussion** section. Each report concludes with (7) a list of **references cited.**

This then is the basic format of the research report. Although this format is not followed by every writer of every paper, it is the style accepted (and expected) by the editors of most biological journals. To be on the safe side, though, verify the format before you begin to write. For a course project, consult your professor. When preparing a paper for submission to a science journal, read the editors' advice to authors, included in each issue of most journals.

The Basic Design of the Research Report

1. Descriptive Title
2. Abstract
3. Introduction
4. Methods and Materials
5. Results
6. Discussion
7. References Cited

If you followed the writing steps in Chapter 5 to prepare your research proposal, you already have a working title (which by now has probably evolved into your descriptive title), a fairly complete introduction, a finished methods and materials section, and your references. Your task now is to prepare figures and tables, organize the results section, revise the introduction, write the discussion, and list the references.

Some Practical Suggestions

Having come this far in your academic program, you have probably developed some methods of composing that you like and that work well enough. Nonetheless, you might try an approach used by many scientists. Assuming that you will be with this project for a sustained period of time, work on each section separately and store it in its own file folder along with references and data relevant to that section. Also try drafting each paragraph on a separate sheet of paper. This makes rearranging paragraphs easier. As you rewrite, some of your paragraphs may grow to a full page or even longer. No problem. Of course, if you do all your writing on a word processor, you can rewrite and reorganize much more quickly. But don't forget to save your material frequently and to run a hard copy at the end of each work session. Of course when you put the material together, you will have to make sure the parts fit smoothly.

The Descriptive Title

The purpose of the title of a research report is to tell the reader concisely and accurately what your paper is all about — that is, what you did and what material you worked with. A good title lets readers decide whether a paper is pertinent to their interests and fields. Consider the titles of the papers you read for this report. How good were they according to these criteria? Could you write a better title for any of them?

Let's take a look at some students' titles. Can you see what's wrong with each? How would you change them?

1. Autoplastic Blastemal Transplantations in *Notophalmus viridescens.*
2. A Summary of Cetological Study.

The first title fails to include the purpose of the research and the common name of the animal used. The second title fails to tell whether this is a research paper or a monograph reviewing whale studies.

The Title Page

Although in some cases you may be asked to place the title at the top of the first page of the report itself, the more common practice is to have a separate title page. Figure 6.1 shows an example.

One last point on titles — just when you think you have one you like, a brainstorm may hit and you will want to change the title. Don't worry. Apart from the abstract, the title is frequently the last thing finished on a research report. Never feel you can't change anything just because it is already typed. If a change would improve the paper, make it.

The Abstract

Although the abstract goes second in the report, it is generally the last part to be written. Because it is a summary of the paper, you need the completed paper in order to write it. Brevity is blessed in

HISTOLOGICAL OBSERVATIONS OF DERMAL
PAD CHANGES DURING GROWTH HORMONE
SUPPORT OF LIMB REGENERATION IN THE
HYPOPHYSECTOMIZED NEWT, TRITURUS
(NOTOPHTHALMUS, DIEMICTYLUS) VIRIDESCENS

by

Jean Dombroski

A THESIS SUBMITTED IN PARTIAL
FULFILLMENT OF THE REQUIREMENTS
FOR THE DEGREE OF
BACHELOR OF ARTS
IN
BIOLOGY

ST. MICHAEL'S COLLEGE
1979

Figure 6.1 *Title page.*

an abstract — usually 250 words is the limit. But you're expected to tell the reader why you did the experiment, how you conducted the experiment, what your results were, and finally the significance of those results. In other words, the abstract should explain all the major points of the report and be understandable without reference to the report itself. All in 250 words or less. It isn't easy, but it is important. A busy reader will usually scan a paper's title and abstract to see if the report is worth reading. You make your case here or not at all for many readers.

> WRITING 6.1: WRITING AN ABSTRACT. Go to the library and select a journal that contains abstracts for its reports. Choose an article of interest to you and, without reading the abstract, read the report and write an abstract for it. Then compare your version with the author's. Don't be surprised if yours is better!

The Introduction

This part of the research report states your hypothesis and places your work in the context of research similar to yours. If you were fortunate enough to have written a research proposal before you began your experiments, then much of the work of writing this section has been done already. One change is essential, though: whereas your proposal was written in the future tense, the report should be in the past tense.

The introduction may have three parts: (1) a summary of your project, your hypothesis, and your conclusions, (2) a historical review of the pertinent research, and (3) a brief summary of your experiment as a transition to the next section, methods and materials. Let's look at these parts one at a time.

1. A summary of the project, the hypothesis, and your conclusions. Until a few years ago the results and conclusions were never mentioned until the end of the report. Today it is accepted practice to begin the introduction with a paragraph outlining them. Thus the reader is not kept in suspense waiting to learn "who dunnit."

2. A historical review of the pertinent research. This section of the introduction justifies your particular study by placing it in the context of earlier work. It is not necessary to detail each individual study; summarizing the general state of knowledge and discussing a few pertinent papers in detail is sufficient. Your ability to synthesize concepts based on many studies and to select the most important is the key to learning to write an effective introduction. The following paragraphs from an actual senior research paper (Dombroski 1979) illustrate the principle.

> The phenomenon of newt regeneration has been observed and investigated for over one hundred years. From the vast amount of research that has accumulated, three conclusions have been made as to the requirements for the occurrence of this process. These conclusions are "(a) limb tissues must be injured, (b) a wound epidermis, free of underlying dermis, must cover the amputation surface, and (c) nerves must be present in sufficient quantity at the level of amputation" (Tassava and Mescher 1975). All three of these conditions must be met in order for normal limb regeneration to occur.
>
> The importance of the wound epidermis has been pointed out by numerous investigators (Goss 1969; Singer and Salpeter 1961; Mescher 1976; Thornton 1957, 1968; Trampusch and Harrebomee 1965). The epidermis and the underlying cells are in intimate cell-to-cell contact (Salpeter and Singer 1960; Schmidt 1962). This contact allows for intercellular communication which seems necessary for regeneration to occur. The mitotic activity of the epidermis increases (Chalkley 1954) and the formation of an apical cap over amputation surfaces results.

> The apical cap is necessary for regeneration to occur (Goss 1956; Thornton 1957, 1958). Godlewski (1928) attempted to transplant skin over the amputation surface to inhibit regeneration but was unsuccessful, for the limbs regenerated at the edges of the transplant where there was intimate contact between wound epithelium and the underlying tissues.

After this beginning Dombroski continues to summarize those studies whose results have a more direct bearing on her work.

3. A brief summary of your experiment and your conclusions as a transition to the next section, methods and materials. If the introduction is several pages long, it's a good idea to conclude with a connection to the methods and materials section. Here is an example of a closing paragraph (Dombroski 1979).

> Schotte and Hall (1951) observed thick dermal pad formation fifteen days post-hypophysectomy. The growth hormone affected the dermis in some way which then allowed for normal regeneration to occur. The effect of this hormone on the dermis is still under investigation. The purpose of this investigation is to determine the early histological effects of a bovine growth hormone preparation on the dermal pad.

The author concludes the paragraph by stating the purpose of the experiment, but not the outcome. She forgot to use the past tense in stating the purpose of the study. Adding one more sentence would make this a complete closing statement for the introduction: "Results indicate that the effect appears to be the prevention of dermal pad formation."

WRITING 6.2: WRITING THE INTRODUCTION. If you wrote a complete introduction for your research proposal, all you need do now is write parts 1 and 3, the introductory paragraph and the transition paragraph, and change the verb tense to past. If you didn't write a proposal earlier, you'll need to start from scratch; see Introduction, Chapter 5. Whether you are revising your proposal's introduction or drafting a new one, give some thought to the organizational pattern of the literature review. Should you summarize the topic in a chronological sequence or would it be better to group opposing viewpoints? Use whichever method suits your subject better.

Methods and Materials

Whenever scientists present the results of their research, they also tell their colleagues how to replicate the experiment that yielded those results. The methods and materials section provides that information.

If this section was part of your original research proposal, all you need do now is review it for clarity and note any changes you introduced during the actual experiment stage. However, if you did not write a proposal, you'll need to begin at the beginning. Assume that your readers will have no idea of what you have done. Write a draft in a narrative style as though you are telling someone how to set up the experiment. Try to picture another student assembling the materials and going through the steps. When you finish this draft, you may have more information than is necessary, but everything essential should be there. Review the draft: does it include a description of the material you used? are your sources listed? are all details clearly presented? is there too much or too little information?

If you used any standard methods of analysis or followed a well-known procedure, you do not have to repeat all the details. Simply reference the sources. In ecological studies, for instance, many of the techniques for both collection and analysis are given in the American Public Health Association's (APHA) publication, *Standard Methods for the Examination of Water and Wastewater.* Knowledgeable authors need only refer to a particular technique, adding a reference such as (APHA 1985). Should you need to modify a particular standard method, make that clear ("according to Smith [1981] with the following modifications") and describe your changes.

The methods and materials section for Dombroski's paper (1979) ran to two and a quarter pages with one page devoted to a table illustrating her schedule for sacrificing animals.

METHODS AND MATERIALS

Triturus (Notophthalmus Diemictylus) viridescens were obtained from Connecticut Valley Biological Supply Company. They were placed in glass stacking dishes half-filled with water and maintained at 22 ± 1°C on a 12 hours light/dark cycle.

Fifty newts were hypophysectomized by anesthetizing them in thirty percent saturated tri-chlorobutanol. A transverse incision was made through the parasphenoid bones, anterior to the visible pituitary gland. This was followed by a second incision, perpendicular to the first. The resulting flaps of tissue were reflected backwards, exposing the pituitary gland, which was then removed using watchmaker's forceps. The reflected tissue was returned to an approximation of its original position. The newts were placed back into water, which was changed every other day following hypophysectomy. They were not fed before or after the procedure. After a delay of five days to allow the metabolism of any residual hormone(s), the right and left forelimbs of all newts were amputated through the radius and ulna. Twenty days post-hypophysectomy

TABLE 1

Schedule for sacrificing newts following growth hormone therapy initiation.

	No. of newts	hypophysectomy	amputation	Growth hormone therapy (GHT)	Control and Experimental Newts Sacrificed	
					post amputation	post GHT
Group 1	6	day 0	day 5	day 20	day 15	3 hours
Group 2	6	day 0	day 5	day 20	day 15	6 hours
Group 3	6	day 0	day 5	day 20	day 15	12 hours
Group 4	6	day 0	day 5	day 20	day 16	24 hours
Group 5	6	day 0	day 5	day 20	day 18	48 hours
Group 6	6	day 0	day 5	day 20	day 20	96 hours
Group 7	6	day 0	day 5	day 20	day 24	8 days
Group 8	6	day 0	day 5	day 20	day 32	16 days

(fifteen days postamputation), the newts were randomly divided into experimental (thirty-two newts) and control (sixteen newts) groups. The experimental newts received injections of 0.2 mg of bovine growth hormone (Calbiochem) in 0.05 ml of distilled water per alternate day. Control newts received injections of 0.05 ml of distilled water. Injections were administered intraperitoneally with a 26 gauge needle in a oblique fashion to prevent leakage.

Four experimental newts and two control newts were sacrificed at different time intervals after growth hormone therapy began, according to the following schedule (Table 1).

The limb stumps and heads that were removed were fixed in ten percent neutral buffered formalin, decalcified with seven percent EDTA in fixative, embedded in paraffin, sectioned at twelve microns, and stained using Masson's Trichrome stain (modified).

WRITING 6.3: DRAFTING THE METHODS AND MATERIALS SECTION. Write a first draft of your methods and materials section. You might like to pattern this after the one used by the author of an article you read for your research. Otherwise, follow the advice given earlier in this chapter and in Chapter 5. By all means ask someone to read and comment on your draft, then revise accordingly.

Results

Before this section is written, you will have completed your entire experiment and performed all the statistical analyses required. If you are reading this chapter before you actually start your experiment, congratulations! Your foresight can save you time and lead to a better report. Many students design experiments and proceed to collect mounds of data without regard to how the information will be presented in the final paper. They know, in the back of their minds, that they will probably have figures and tables to represent all this information. They assume that the actual design of these figures and tables takes place during the drafting of the report. Not so! If you have planned ahead, you can construct the tables and figures as the data is collected. This is not always possible, but when it is, it makes your work easier.

Tables and Figures

Tables are lists of numbers and **figures** are diagrams or illustrations. You'll find a sample table on page 102 and a sample figure on page 106. Neither includes every single bit of information collected; they show representative examples. Tables tend to have more specific information than figures but are harder to read because of the great detail. Figures generally show average results, usually with some reference to standard deviations from the averages. For more help in constructing your tables and figures, we recommend that you study one of the many good references available such as Moroney (1968).

Every table and figure must have a title. This should explain the material without reference to the text. The title of a table appears above the table. Titles of figures are placed below. Tables and figures are numbered separately and consecutively as they appear in the paper. You might have ten tables and five figures, for instance. The tables would be numbered 1 through 10 and the figures 1 through 5. The table or figure should appear in the paper as soon after you discuss it as possible. Break this rule only when your professor or editor requests that all figures and tables be presented in an appendix.

If you want to use a table or figure from an outside source, place it wherever appropriate in your text. Number it according to its sequence in the paper and cite your source in parentheses after the title.

Most figures, whether in published articles or in student pa-

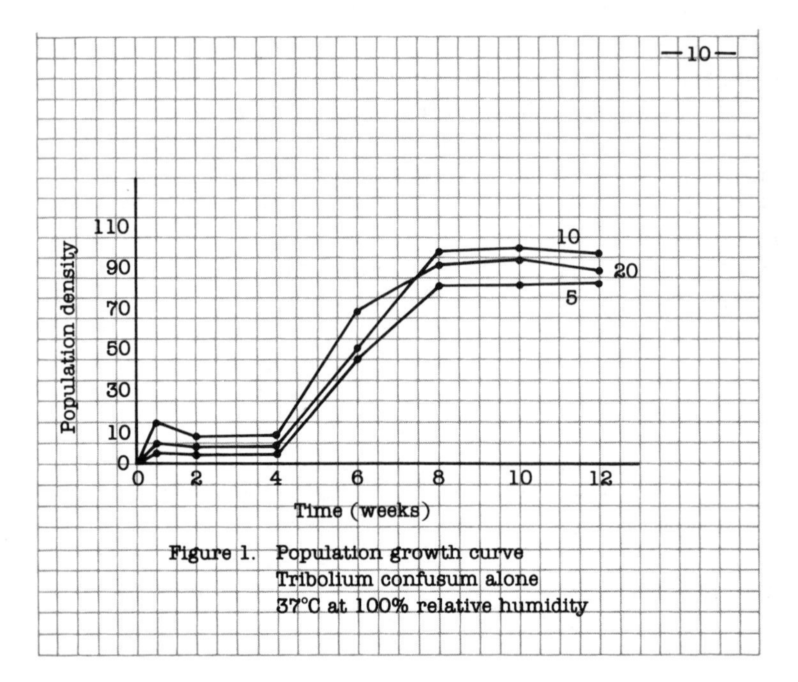

Figure 6.2 *Graph from a general biology student's report.*

pers, contain only a few lines or bars. If you produce figures with more than three or four lines, the potential reader confusion is more of a hardship than if you had presented two or three figures to illustrate your points.

Graphs are special kinds of figures. Line graphs should be used only for continuous variables like temperature or hours of daylight. Bar graphs are used to illustrate data which is discontinuous, such as readings from different areas (e.g., sample points along a river), at different times (e.g., weekly interval sampling), of different items (CO_2 *vs.* O_2). Line graphs are preferable to bar graphs, which are preferable to tables. When tables are used, the information they provide should be summarized in a graph. Figure 6.2 shows a graph that a student included in her Results. Can you read it and understand its significance? How might you improve it?

Sometimes the nature of your results cannot be presented in a table or a figure. Dombroski's results included drawings to illustrate the changes she observed. Other researchers might present their results through photographs. Whatever your means of presentation, the outcome should be clear and easily understood.

Narrative

When your data have been organized into tables and figures, you must prepare a narrative to accompany them. Keep your comments brief and limited to pointing out specific areas for later explanation in the discussion section. Here is Dombroski's treatment of one of her figures:

> The amputated limb indicated in Figure 1 is a control which the experimental groups were compared to. This limb is from a control newt sacrificed twenty-four hours post-GHT initiation. It shows the thick dermal pad beneath the epidermis, a characteristic of a non-regenerate.

With the completion of your results section, you can turn your attention to the final and most important part of your paper — the discussion.

Discussion

In this section you review the results of your experiment and compare them with those of the previous investigators you wrote about in the introduction. Take each important point made in your results section and explain its significance, commenting as to whether or not your results support your hypothesis and expectations and discussing them in the light of prior research.

Occasionally it may be necessary to introduce new material from the literature while comparing results. However, if the new reference is of sufficient importance to be discussed now, it probably should have been included in the introduction. Consider adding this reference to your introduction now.

After reviewing the results, explaining their significance, and comparing them with previous works, complete the discussion with a statement of the impact of your results on the total body of knowledge addressed by this study. You may also wish to point out new areas for research that you discovered.

Here is how Dombroski concluded her discussion section:

> When regeneration begins to take place, the cells which are believed to contribute to the blastema cell population have already accumulated in the dermal pad, potentially able to undergo dedifferentiation. A possible model for regeneration in the case of the hypophysectomized newt is proposed. If growth hormone increases collagenase synthesis, the intercellular matrix would be broken down, leaving the epidermis free from dermis, available for innervation by advancing nerves. The nerves would then release their trophic factor (FGF), resulting in proliferation of dedifferentiated fibroblasts. This model would explain early blastema formation and general accelerated regeneration with the onset of GHT.
>
> The above is purely speculation as this investigation did not concern itself with the biochemical or neurogenic factors involved in regeneration. The recent studies have been exciting in their suggestions and it is the possibilities of these suggestions which lend themselves to further research in the complicated and many-faceted area of regeneration.

Some authors like to summarize their whole paper with either a paragraph similar to the abstract or in list form. Such a summary would highlight the impact of your work.

Checklist for the Research Report

1. Descriptive title
   Does the title tell the reader what you did and what material you worked with?
   Does your title page follow the format shown on page 97?
2. Abstract
   Is the abstract 250 words or less?
   Do you tell why and how you did the experiment and what the results and conclusions are?
   Do you use the past tense here and throughout the paper?
3. Introduction
   Have you presented all the important literature?
   Are all references properly cited?
   Have you summarized your project and conclusions?
4. Methods and materials
   Are your methods clear enough to be followed by another researcher?
   Have you used prose rather than tabular form?
   Have you included all necessary citations?
   Have you indicated your statistical methodology, if necessary?
5. Results
   Are tables and figures properly constructed, labeled, and ordered?
   Does the narrative indicate the significance of each table and figure and of the statistical results?
6. Discussion
   Do you review your results and compare them to the literature discussed in the introduction?
   Do you point out areas for new research?
   Do you present your conclusions on the problem?
7. References cited
   Have you listed all the references cited and only those cited?
   Are references in alphabetical order?
   Are the citations complete and accurate?

After you have finished a draft of the discussion section, you can write the abstract (explained earlier in this chapter). That leaves only one section to go — references cited.

References Cited

This section of the research report lists all the literature you actually used in the body of your paper. It should not include any references you read but did not cite. You would list all the sources consulted only if your professor specifically calls for it; then this section would be titled "Bibliography." The purpose of the list of references is to direct the reader to other work on the subject.

The generally accepted style is to list references alphabetically by author's last name. See Chapter 8, Documentation Style Sheet, for complete information on compiling your list of references cited.

Revision

Revision matters. It marks the distinction between the amateur and the pro. Our best advice is that when you have a good draft you reread the revision section of Chapter 4. Then review Chapter 9, A Concise Guide to Usage, and Chapter 10, Make Punctuation Work for You. To reiterate, your research report represents weeks of hard work. Respect that effort and others will too.

[7] *Principles of Library Research and Basic Bibliography*

PREVIEW: *If only you knew what your college library contains! Whenever you have to write a term paper or a research proposal, you turn to the library. In this chapter you'll learn, step by step, just how to find the information you need.*

- *Finding the literature — at home*
  - *Reference cards*
- *Finding the literature — in the library*
  - *Card catalog*
  - *Cataloging systems*
  - *Serials and periodicals*
  - *Indexes and abstracting services*
  - *On-line computer search*
  - *Interlibrary loan*
- *Using the literature: taking notes*
  - *Summary*
  - *Paraphrase*
  - *Quotation*
  - *Photocopying*
- *Useful reference sources in the life sciences*

"The library doesn't have anything in it."
"All the references I wanted to use were checked out (or missing, or cut out)."

You've heard these complaints often enough, maybe even expressed them yourself. Yet these common gripes are rarely accurate. It is the truly unique library that doesn't hold anything of value for the researcher. Even those institutions with limited resources will provide enough material for your research if you know where and how to look.

It really isn't such a big mystery. Some common sense and a little knowledge are all you need. This chapter will show you how to find those materials and build a basic bibliography for your paper, as well as how to read the literature and use it properly.

FINDING THE LITERATURE — AT HOME

Let's say you have prepared a list of possible topics for your paper. Now you are ready to check the availability of sources before deciding on a specific topic. As you do this, you are also compiling a working bibliography — a list of journal articles and books that you will read later on.

Where do you begin? Although the library would seem the logical choice, first consider what other sources are available. Students commonly overlook the most obvious — their textbooks and laboratory manuals. An up-to-date general biology text is an excellent place to begin your search, even if you have to write a specialized paper for an upper-level course. Turn to the index and look for your topic there, then check the treatment of that topic in the text. Some index references may lead you to a passing mention of the topic in the text, others to a more complete treatment. Then turn to the end of the chapter in which the topic appears. There you should find a list of additional readings ranging from journal articles and book chapters to entire books. Add promising references to your working bibliography. If the book's index does not list your topic, don't despair — consult the table of contents. Under which chapter heading does your topic fall — animal behavior, perhaps, or genetics? Check the list of additional readings at the end of those chapters. Many biology textbooks also contain a substantial list of additional readings at the end of the book. See if yours does.

Lab manuals provide another close-to-home source. Look at the end of the experiment section on bio-chemical reactions or plant pigments or whatever else seems close to your topic. You will probably find a list of additional readings.

Reference Cards

Although at this stage of your research you won't be reading the sources and taking notes, you do have to record information for your working bibliography. Researchers do this in several ways, but perhaps the simplest and most reliable is to use a card system. Get a supply of 4″ × 6″ lined index cards. Use one for each reference. Leave the top line blank — you'll fill in the library call number here later. On the second line, write the author's full name, last name first. If an article or book has more than one author, record all names. At the right margin of this line place the publication date. Use the third line for the title of the article or book. For a journal article, put the name of the journal, its volume number, and inclu-

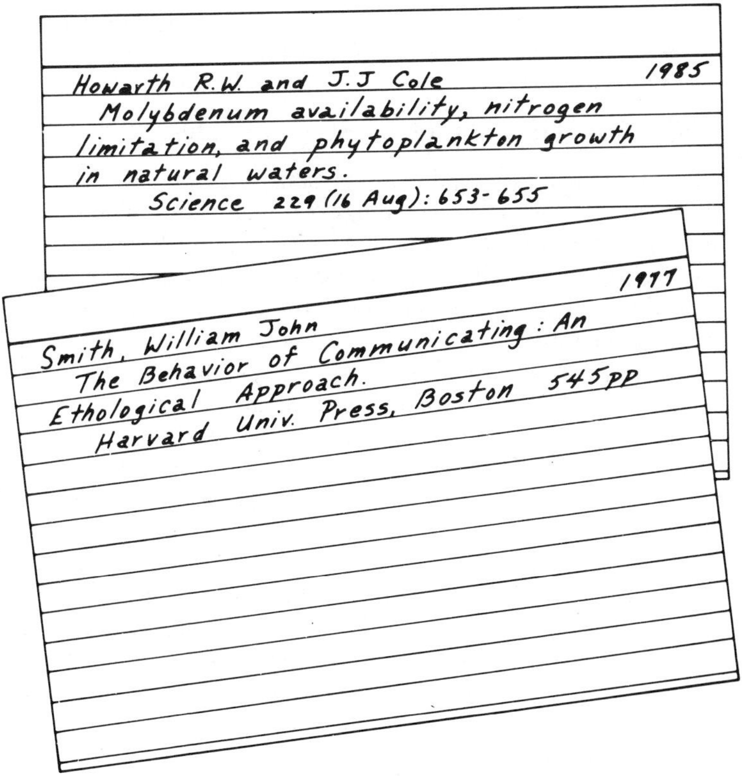

Figure 7.1 *Sample 4 × 6 reference cards. Top — journal article, bottom — book reference.*

sive pages on the fourth line. For a book, list the name of the publisher, the city of publication, and the inclusive pages. The remainder of the card is for your notes when you read the material. See Figure 7.1 on preceding page for sample reference cards.

If you follow these suggestions, you will find that even before you leave your room to go to the library you can have a list of possible sources ready to check for availability.

WRITING 7.1: FINDING THE LITERATURE I. List two possible paper topics for a biology course you are taking. Use your textbook and other books in your room to start a reference list. See how many references you can find for each topic. Use the card system suggested above.

FINDING THE LITERATURE — IN THE LIBRARY

Whether your library represents the small collection of a community college or the massive holdings of a major research university, the remainder of this chapter should be helpful: the advice holds true for libraries of all sizes. Of first importance is learning what your library can provide. The staff of most college libraries offer group orientations or at least a pamphlet describing self-guided tours. Your improved ability to use the resources will more than make up for the hour or so taken by the tour.

Every library, regardless of size, has several resources of value to the researcher. (In smaller libraries some of these functions may be combined in one place.) The **circulation desk** checks books and other materials in and out of the library. The **reserve desk** holds books and periodicals set aside by professors for short-term student use. The **reference desk** is the source of help in finding information as well as the center of a special collection of basic reference materials such as encyclopedias, bibliographies, dictionaries, indexes, even out-of-town telephone books. The **periodicals section** contains current and past issues of popular magazines and scholarly journals. Large libraries also have media centers, map rooms, microfilm collections, manuscript and rare book libraries, and government document centers.

Card Catalog

Your search will probably begin at the card catalog (or the computer terminal, if your library's catalog has been computerized). Each book the library owns is listed here in three ways: by title, by author, and by subject. If you know the title or author of a book you want, simply find the book card, which shows the call number. The **call number** assigned to each book is unique — no other book in the collection has that number. And that number tells you where to find the book on the shelves. (A few libraries have closed stacks, denying students access to the shelves. In a closed-stack library you fill out a call slip and present it to a librarian, who will get the book for you.) If you are not looking for a specific book but want to see what is available on your topic, you need to check the subject catalog. Look up the topic (photosynthesis, for instance) and you will probably find a number of books listed. Make a reference card for each book that looks valuable. If you are not sure whether the book will fit your topic exactly, make a card. It might prove useful if your topic shifts ever so slightly. And don't forget to include the call number.

Useful as the card catalog is, you must realize that it is not a complete listing of all the books in the library that treat your topic. Furthermore, although the titles of journals will be included, titles of articles in those journals will not be. You will have to supplement the card catalog with other resources described later on.

Figure 7.2 on page 116 shows a sample card catalog set of title, author, and subject cards. These would appear in different places in the card catalog, filed alphabetically.

Cataloging Systems

As you take notes at the card catalog, you will copy the call number of each reference. The make-up of that call number will depend on which of two cataloging systems your library uses: Library of Congress or Dewey Decimal. Both are attempts to classify and arrange the holdings of a library and make them accessible to readers. At some point during your research you will want to understand how

QL 776 .S64 The behavior of communication: an ethological approach. — Title card

Smith, William John.
The behavior of communicating: an ethological approach / W. John Smith—Cambridge, Mass.: Harvard University Press, 1977.

QL 776 .S64 ANIMAL COMMUNICATION. — Subject card

Smith, William John.
The behavior of communicating: an ethological approach / W. John Smith—Cambridge, Mass.: Harvard University Press, 1977.

Main Entry card (Author)

author ↓ · *title* ↓

QL
776
.S64

Smith, William John.
The behavior of communicating: an ethological approach / W. John Smith—Cambridge, Mass.: Harvard University Press, 1977.

↑ *imprint*

↑ *call number*

viii, 545 p. : ill. ; 24 cm.

Bibliography: p. [469]–513. ←— *bibliographic note*
Includes index.
ISBN 0-674-06465-8

↓ *tracings (list of entries)*

1. Animal communication. I. Title.
QL776.S64 591.5′9 76-43021
MARC

Library of Congress 76

Figure 7.2 *Card catalog set.*

your library's system works. The information below will let you check the shelves for holdings in your research area.

The **Library of Congress system** uses the alphabet for the divisions: general science books under **Q,** for instance; medicine under **R;** agriculture and forestry under **S;** and engineering and technology under **T.** This system allows for finer classification by the addition of a second letter followed by numbers. **QH 351,** for example, is the classification for morphology. The Library of Congress listings for the biological sciences follow:

Library of Congress Classification System for the Biological Sciences

QH 72	Nature Conservation. Landscape Protection
QH 90	Water. Aquatic Biology
QH 91–95	Marine Biology
QH 96	Freshwater Biology
QH 201	Microscopy
QH 301–349	General Biology
QH 351	Morphology
QH 425	Animals
QH 426–470	Genetics
QH 471–499	Reproduction
QH 506	Molecular Biology
QH 515	Photobiology
QH 540–559	Ecology
QH 573–705	Cytology
QK 1–102	Botany
QK 110–195	North American Botany
QK 475	Trees and Shrubs
QK 494	Gymnosperms
QK 495	Angiosperms
QK 520	Ferns
QK 534	Mosses
QK 564–580	Algae
QK 581	Lichens
QK 600–638	Fungi
QK 641–707	Plant Anatomy
QK 710–929	Plant Physiology

QK 930–935	Physiographic regions. Water.
QK 936–939	Physiographic regions. Land.
QK 940–977	Physiographic regions. Topographic divisions.
QL 1–	Zoology
QL 362–599	Invertebrates
QL 605	Vertebrates
QL 614–638	Fishes
QL 667	Amphibians
QL 671	Birds
QL 700–739	Mammals
QL 750	Animal Behavior
QL 799	Morphology, Anatomy
QL 951	Embryology
QM 1–	Human Anatomy
QM 111–131	Skeleton
QM 178	Vascular system
QM 301	Organ of Digestion
QM 451	Nervous System
QM 501	Sense Organs
QM 531	Regional Anatomy
QM 550	Human and Comparative Histology
QM 601	Human Embryology
QP 1—	Physiology
QP 88	Physiology of Tissue
QP 99	Blood
QP 101	Cardiovascular system
QP 141	Nutrition
QP 186	Glands
QP 301	Movements

QP 351–495	Neurophysiology and neuropsychology
QP 501–801	Animal Biochemistry
QP 901	Experimental Pharmacology
QR 1–74	Microbiology
QR 75–200	Bacteria
QR 201	Pathogenic Micro-organisms
QR 355–484	Virology

The **Dewey decimal system** identifies the main divisions by numbers. The Pure Sciences are listed under 500 to 599. The Applied Sciences, including Medicine, are listed in the 600 to 699 range.

Dewey Decimal Classification System for the Biological Sciences

500	Science in General
503	Dictionaries, Encyclopedias
505	Periodicals, Magazines, Reviews
507	Education
509	History of Science
510	Mathematics
520	Astronomy
530	Physics
540	Chemistry
550	Geology
560	Paleontology
570	Biology, Archeology
573	Natural History of Man
574	Physiology and Structural Biology
575	Evolution
576	Origin and Beginning of Life
577	Properties of Living Water

578	Microscopy
579	Collectors Manuals
580	Botany
581	Physiology and Structural Botany
590	Zoology
591	Physiologic Zoology
592	Invertebrates
596	Vertebrates
610	Medicine
614	Public Health
620	Engineering
630	Agriculture

Serials and Periodicals

Serials are printed materials that are issued on a regular, recurring basis. They may appear as often as the daily newspaper or as infrequently as an annual monograph. Those not published daily are called periodicals. A primary appeal of serials and periodicals to the researcher is their frequent publication, which allows almost constant updating of information. We are living in the midst of an "information explosion" — knowledge is created faster than most of us can follow. A six-year-old book on genetics may be out of date by the time you find it on the library shelves. Researchers, then, must supplement their reading in books by checking the subject in more recent journals.

The treatment of serials varies from library to library. Some libraries house their current journals in a readily accessible room or section, filed alphabetically by title. Other libraries catalog their journals and maintain title cards for them in the card catalog. Each year the library binds back-issues of a journal into book form. In some libraries these bound journals are protected by a Periodicals Librarian in a restricted area; you ask for an article by author, title, journal, and year. In other libraries bound journals are kept on open shelves.

WRITING 7.2: SERIALS AND PERIODICALS. When researching for a paper, it is helpful to know which biological journals your library carries. Near the end of this chapter you will find a listing of major informal journals, review journals, and research journals. Find the call numbers of those biology and general science journals carried by your library. (Call numbers are listed in the card catalog and in a printout at the reference or periodicals desk.)

Indexes and Abstracting Services

Bio-researchers can save a good deal of time by learning to use the special aids available to them in the form of indexes and abstracting services. *Current Contents* is a monthly index that reproduces the tables of contents of major journals. It provides a quick way to review current titles but doesn't give additional information about the articles. *Biological Abstracts,* on the other hand, includes an abstract of each article. This useful source is divided into a variety of sections that cover the fields of bioscience from aerospace biology to virology. Even if some of the articles abstracted appear in journals not available to you, the information gained from the abstract can be used and *Biological Abstracts* themselves cited as the source. Issues appear semi-monthly and, in addition to the abstracts, contain five indexes: author, biosystematic, generic, cross, and subject. A semi-annual index is also available.

Figure 7.3 on page 122 shows portions of several pages of *Biological Abstracts;* items a and b are from the subject index and item c shows a sample abstract. Let's see how the search process works. The working title of your project is "Genetic control of the immune response," and you are just beginning your search. Your working title has two good key words: *genetic* and *immune*. By checking these key words against the index, you find the two partial titles shown in Figure 7.3 (a and b). The first is listed under the heading IMMUNE and the other under GENETIC. Both carry the same abstract number, 13950. When you refer to that number in the Abstracts section of this volume, you find the complete reference abstract (shown in c). Take note of all the information available to you with the abstract.

Biol Abstr 81(2):SI-133 (a)

Subject Context	▼ Keyword	Ref. No.
TANCE OF DNA ANTI-DNA	IMMUNE COMPLEXES HUMAN SYSTEMIC L	14075
ODIES AND CIRCULATING	COMPLEXES LEVEL IMMUNOGLO	15875
ONIAS ANTICOMPLEMENT	COMPLEXES PROGNOSIS/ CIRCU	14097
ED BY HEART-AFFECTING	COMPLEXES PUMP IMPAIRMENT	16503
EMOVAL OF CIRCULATING	COMPLEXES QUANTITATIVE ISOL	13978
N/ THE CORRELATION OF	COMPLEXES WITH ACUTE PHASE	14141
E MARROW IN ACQUIRED	DEFICIENCY SYNDROME A HISTO	14161
VIRUS HUMAN ACQUIRED	DEFICIENCY SYNDROME ANTIGE	19528
AR LAVAGE IN ACQUIRED	DEFICIENCY SYNDROME HUMAN	16396
RACELLULARE ACQUIRED	DEFICIENCY SYNDROME LOWENS	14851
ATIENTS WITH ACQUIRED	DEFICIENCY SYNDROME OR ACQ	14968
G FRAMES OF ACQUIRED	DEFICIENCY SYNDROME RETROV	19527
S HEMOPHILIA ACQUIRED	DEFICIENCY SYNDROME TRANSM	18141
VIRUS TYPE III ACQUIRED	DEFICIENCY SYNDROME VIRUS C	19548
RECOVERY OF ACQUIRED	DEFICIENCY SYNDROME-ASSOCI	14968
SARCOMA OR ACQUIRED	DEFICIENCY SYNDROME-RELATE	11432
YNDROME OR ACQUIRED	DEFICIENCY SYNDROME-RELATE	14968
-RETICULAR STRUCTURES	DEFICIENCY/ ACUTE GLANDULAR	14970
YTE ANTIBODIES HUMAN	DISORDER GRANULOCYTE IMMU	14158
AMINASE CELL MEDIATED	DYSFUNCTION/ DIFFERENTIAL IN	14028
IX PROTEIN DETECTED BY	FLUORESCENCE WITH MONOCLO	14938
RAFT-VS.-HOST ALTERED	FUNCTION SISTER CHROMATID E	13441
TIS B E ANTIGEN VACCINE	GLOBULIN ANTIBODY RESPONSE	13944
MUNOLOGIC PROCESSES	HOMEOSTASIS ENDOCRINE SYST	15983
BOCYTOPENIC PURPURA	IMBALANCE SEX DIFFERENCE/ DI	10630
CLASS I HLA ANTIGENS IN	INTERFERON-TREATED HUMAN	14139
S BY DNA RECOMBINANT	INTERFERON/ IMMUNOCHEMICA	15539
CYTE PROGENITOR CELL	MECHANISM/ IN-VITRO TYPING O	10627
ATHOGENESIS FIBROSIS/	MECHANISMS FOR HEPATIC FIBR	14202
CTS RAT INFLAMMATION	REACTION/ LONG-TERM CULTUR	18868
VATOR BLOOD SAMPLING	REACTIVITY COAGULOPATHY PR	16816
PPRESSOR MECHANISMS	REGULATORY DYSFUNCTION PH	14038
SIS HISTOCOMPATIBILITY	REJECTION/ AVIAN SPINAL CORD	14009
NINE I. ENHANCEMENT OF	RESPONSE BY SWAINSONINE IN-	16823
N RATS T CELL MEDIATED	RESPONSE CASEIN/ INHIBITORY	19418
FT REJECTION HUMORAL	RESPONSE CELLULAR IMMUNE R	14107
N-ANTIBODY COMPLEXES	RESPONSE EFFECTOR ARM/ SUP	14058
EXPRESSION OF MURINE	RESPONSE GENES GAMMA INTER	14105
ULATION OF THE HUMAN	RESPONSE HEPATITIS B SURFAC	13943
LIC-ACID ON THE MURINE	RESPONSE IMMUNOLOGIC-DRUG	14107
SSION OF THE HUMORAL	RESPONSE IN MICE LEUKOCYTE	19245
ED TUMOR GROWTH AND	RESPONSE IN RATS NATURAL KIL	15402
N HEMOLYMPH PROTEIN/	RESPONSE OF HOLOTRICHIA-OBL	14410
OID THERAPY/ SYSTEMIC	RESPONSE OF PATIENTS WITH A	14191
R LYMPHOCYTES B CELL	RESPONSE RHEUMATOID ARTHRI	13779
OF BOTULINUM TOXIN ON	RESPONSE STAPHYLOCOCCUS S	14881
E LYMPHOKINE SOLUBLE	RESPONSE SUPPRESSOR DURIN	14253
RO DIROFILARIA-IMMITIS	RESPONSE THIRD-STAGE LARVA	16382
RAPEUTIC EFFICACY/ THE	RESPONSE TO A SCHISTOSOMAC	11483
OPROTEIN GC/ HUMORAL	RESPONSE TO HERPES SIMPLEX	13949
E/ GENETIC CONTROL OF	RESPONSE TO SALMONELLA-ENT	13470
LIN ALLOTYPES AND THE	RESPONSE TO WHEAT GLIADIN I	13673
LASMS IN MICE CELLULAR	RESPONSE TUMOR TEXTURE IM	15416
EPRODUCTIVE FUNCTION	RESPONSE-ASSOCIATED GENE/	13407
MYELITIS INFLAMMATION	RESPONSE/ DOSE-DEPENDENCY	15885
HUMAN T CELL HUMORAL	RESPONSE/ MONOCLONAL ANTI	14237
ULATING FACTOR MOUSE	RESPONSE/ STIMULATION OF BO	14242
NE RESPONSE CELLULAR	RESPONSE/ THE ADJUVANT EFFE	14107
SPECIFIC CELL-MEDIATED	RESPONSES AFTER INFECTION O	[illegible]
L ENV GENE PROTECTIVE	RESPONSES HLA VACCINE DESIG	13950
ESPONSE AND CELLULAR	RESPONSES IN A MURINE MODEL	14177

GENETIC - GESTATIONAL (b)

Subject Context	▼ Keyword	Ref. No.
ES HLA VACCINE DESIGN/	GENETIC RESTRICTION OF IMMUNE RESP	13950
GONIA CANCER PATIENT	RISK/ LONG-TERM INFERTILITY	[illegible]
DROGENASE DEFICIENCY	SCREENING/ THE LABORATORY	10606
NAMIBIA/ BIOCHEMICAL	STOCK STRUCTURE OF THE SOU	13399
POPULATION EVOLUTION	STRUCTURE MODEL/ GEOGRAP	13698
T ALLELE SEGREGATION/	STUDIES OF GAMMARUS IV. SELE	13534
GHT STRUCTURAL GENES	STUDIES/ CHROMOSOMAL CON	13763
RPHOLOGICAL VARIANTS	STUDY/ ISOLATION OF SPORELE	13323
RSITY GENE ACTIVATION	SUBUNIT POSTPROTHORAX BIT	13423
IC AUTOIMMUNE DISEASE	SUSCEPTIBILITY COLLAGEN DEP	14019
TAL MEDICAL PROBLEMS	SYNDROMES/ MOTHERS' BELIEF	17882
/ SELECTION IN COMPLEX	SYSTEMS VI. EQUILIBRIUM PROP	13580
NTIAL GENE ACTIVATION	TARGET MUTAGEN/ INDUCTION	15310
ENICITY GENOTOXICITY/	TOXICITY OF SEVERAL ANTIHYPE	15280
ORMATION FREQUENCY/	TRANSFORMATION OF SOMATIC	13389
GAN HORSES CROSSINGS	TYPES GENERATION DIFFERENC	10012
NADH DEHYDROGENASE	VARIABILITY DIFFERENTIATION/	10948
POPULATION GENETICS/	VARIABILITY IN NORWEGIAN SE	13528
ON POLYPHAGIC HABITS/	VARIABILITY IN POPULATIONS O	13396
DING COMBINING ABILITY	VARIABILITY REGRESSION ANAL	13862
G FOR THE PRESENCE OF	VARIANCE IN FACTORS OF FACE	13696
HOSPHOTRANSFERASE IN	VARIANTS ALPHA METHYLMANN	13605
NOGASTER NONADDITIVE	VARIATION CHROMOSOME REAR	13442
CTERS OF SUGARCANE 1.	VARIATION GENETIC CORRELATI	9955
L VARIABLE LIFE HISTORY	VARIATION LARVAL DEVELOPME	14754
SUGAR UTILIZATION DIET	VARIATION/ ANTIBIOTIC RESIST	14892
YTHROCYTE INBREEDING	VARIATION/ FREQUENCY OF POL	13435
MATHEMATICAL MODEL/	GENETICAL EVOLUTIONARILY STABLE STR	13519
MATHEMATICAL MODEL/	EVOLUTIONARILY STABLE STR	13523
SMIC MALE STERILITY-RF	SYSTEM IN SUNFLOWER HELIA	9940
VER CYST FORMATION/ A	GENETICALLY DETERMINED MURINE MOD	19434
ECIPROCAL CROSSES OF	LEAN AND OBESE SWINE AL	16255
CALLY OBESE OB-OB AND	LEAN MICE THERMOREGUL	10239
EFERENCE BEHAVIOR OF	OBESE OB-OB AND GENETI	10239
ND HYPERINSULINEMIA IN	OBESE ZUCKER RATS HOR	16708
EIN ANTIGENICALLY AND	RELATED TO FACTOR-VII AL	14164
O CONDITIONED CELLS IS	RESTRICTED AT THE ENDO	14110
GESTERONE OVULATION	GENETICS/CHARACTERISTICS OF THE AT	10024
ROGENEITY POPULATION	/GENETIC VARIABILITY IN NOR	13528
METAPHASE MOLECULAR	/THE ORGANIZATION OF TWO	13621
TIME GIBBERELLIC-ACID	CROP INDUSTRY/ ORIGIN OF G	13886
RHIZOBIUM POPULATION	ENZYME POLYMORPHISM IN IS	13330
ROMATOGRAPHY HUMAN	EPIDEMIOLOGICAL STUDY/ HE	10625
E SYMBOLS IN TEACHING	HUMAN WILD TYPE DOMINANC	13291
ED PREDICTION METHOD	MATHEMATICAL MODEL/ BASIC	10021
HYLENE GLYCOL HYBRID	NUCLEAR LOCALIZATION CROS	13707
OD FLUKE/ POPULATION	OF BIOMPHALARIA-STRAMINEA	13429
ORDER LINKAGE GROUP/	OF GRASSHOPPERS 2. THE INH	13426
YTOPLASMIC DIVERSITY/	OF INBRED DROSOPHILA-MELA	13521
ON HYBRID DYSGENESIS/	OF INBRED DROSOPHILA-MELA	13526
NANCE OF SPECIFICITIES/	OF INBRED DROSOPHILA-MELA	13532
RALIA/ THE ECOLOGICAL	OF INTRODUCED POPULATIONS	13517
N SWEDEN/ POPULATION	OF THE PARTHENOGENETIC GA	13449
HOLINERGIC DEFICIENCY	PARIETAL LOBE COGNITIVE SKI	15944
C BARRIER FOSSIL CORAL	PELAGIC DISTRIBUTION VICARI	11605
USE JERVINE TERATOGEN	PROTEOGLYCANS STEM CELL/	11904
RHEUMATOID ARTHRITIS	SKIN ERUPTION PROTEINURIA/	19329
MBE CELL WALL OSMOSIS	SORBITOL/ PREPARATION AND	13710
DULANS FLUOROCHROME	VEGETATIVE STATE/ ESTIMATI	13755

(c)

13950. NEURATH, A. ROBERT*, STEPHEN B. H. KENT, NATHAN STRICK, DENNIS STARK and PHYLLIS SPROUL. (Lindsley F. Kimball Res. Inst., N.Y. Blood Cent., 310 E. 67th St., New York, N.Y. 10021.) J MED VIROL 17(2): 119–126. 1985. **Genetic restriction of immune responsiveness to synthetic peptides corresponding to sequences in the pre–S region of the hepatitis B virus envelope gene.**—Proteins of the HBV envelope (env) are coded for by two adjacent regions of the HBV env gene: the pre–S and S regions. Antigenic determinants corresponding to amino acid sequences of both regions are recognized by human antibodies and are important in virus–neutralizing responses. Protective immune responses to HBV appear to be linked to the major HLA histocompatibility complex. Inbred and congenic strains of mice represent a model system relevant for studies on the genetic control of immune responsiveness of humans to HBV envelope proteins. Such mouse strains were ranked according to their antibody response to the S protein and divided into high [d,q], intermediate [a,k,b], and low [s] responders (letters in brackets indicate H–2 haplotype.) Selected pre–S antigenic determinants can be mimicked with high fidelity by synthetic peptide analogues that are immunogenic without any carriers. Thus it is possible to study directly the genetic control of immune responsiveness to pre–S epitopes mimicked by these peptides without having to consider the influence of carriers or of S protein. The results presented here show that inbred mouse strains can be ranked according to their antibody responses to the synthetic peptide pre–S(120–145) as follows: A/J[a] $\simeq$ SWR/J[q] > C57BL/6J[b] $\simeq$ AKR/J[k] $\simeq$ SJL/J[s] > DBA/2J[d] > BALB/cJ[d]. Only

Here are a few suggestions to keep in mind as you use the abstract indexes. Articles on some topics tend to appear in bursts. One year may see many articles published; the next few years no one seems interested. For that reason you need patience while doing a literature search. Start with the most recent volumes and carefully work your way back through the issues. Save time by going completely through an Index and collecting all the numbers of abstracts before you go to the Abstracts section to read any of them. With a little practice you will soon be able to find appropriate key words for your searches. But be careful: these words sometimes change from year to year depending on shifts in the terminology of a particular research area. A few years ago, for instance, no one had heard of *Undulipodium,* a new term advanced by Lynn Margulis of Tufts University to distinguish the flagella of eukaryote cells from that of prokaryote cells. And a last bit of advice — once you have found a few good recent articles, perhaps even a review monograph, you can shift your efforts to using the references cited in the articles to continue to build your bibliography.

Because of the abundance of current research and publications *Biological Abstracts* is delayed in compiling and publishing their information. To compensate for this time lag, the same publishers provide *BioResearch Index.* While not as complete and useful as the *Abstracts,* the listings are much more up-to-date. Finding an article of interest in any of these sources, you can go to the journal and read the entire paper.

WRITING 7.3: USING *BIOLOGICAL ABSTRACTS.* Find *Biological Abstracts* in your library. Using the subject index locate four articles pertinent to your research. Make out a reference card for each to add to your working bibliography. Ask your librarian for help if the directions given in the front of the issues are confusing.

Figure 7.3 *(facing page) Sample subject index pages and abstract from* Biological Abstracts, *Volume 81 (2) — a and* b *show the two subject entries for Abstract Number 13950,* c *is the actual abstract. Note the information available from the abstract. (From* Biol. Abst. *81(2), January 15 1986. © Used with permission.)*

On-Line Computer Search

So far in this chapter we have described a wide variety of traditional resources to use in building a working bibliography: lists for further reading in textbooks, the card catalog, periodical indexes, and abstracts. The most modern way to do a literature search, though, is by computer. You ask for a specific type of reference and the computer searches all the literature available to it. Then it shows you everything it has listed on that topic. Depending on the capabilities of the service, you may request a reference list, abstracts of the articles you want, or complete copies of the articles. All this can be done in minutes. Several computer search services such as *Dialog* and *Bibliographic Retrieval Services* offer speedy and almost effortless literature searches. But there's a catch, several in fact. First, you have to tell the service what to look for — that is, define the key words that the computer will scan for. If your key words are too broad, you'll get pages and pages of printout, with only a few relevant references. If your key words are too narrow, the results will be good, but so restricted as to leave you ignorant of the larger picture. A second drawback to computer searches is that all the literature may not have been catalogued by the network you are using, so you'll still have to do a lot of work the old-fashioned way. Yet a third limitation is that many schools restrict the availability of this service to graduate students and faculty. Finally, these searches can be expensive, very expensive.

Obviously, your use of this resource will depend on its availability and your need. Most student papers are more effectively researched the old-fashioned way, without the aid of the computer. This method not only enables you to learn your library's resources, but it is actually more effective if you need the information immediately for a 5- to 10-page paper or are still narrowing your topic. If you are preparing a major term paper or a research paper and have already decided on a specific topic which is limited to the literature of the last ten years (e.g., AIDS), then an on-line search can be very helpful. Such a search can provide a comprehensive survey of the literature or simply identify key articles and authors for you. Whether or not you decide to use a computer search service now, you should have some idea of what your library is prepared to offer you. The reference librarian can tell you what the library has and the policies for use.

Interlibrary Loan

What do you do when your literature search turns up items that are not available in your library? If the missing item is essential to your research and you have left yourself enough time, you can request an interlibrary loan. Depending on this service for too many references can be costly in time and money, though. You don't want to be surprised by a $40 bill for copying a book you were only marginally interested in. If you are thinking about requesting material, check your library's policies. To fill out an order, you will need all the information from your note cards: author, book title or article title and journal, date of publication, and page numbers. You may need your professor's approval as well. In two to three weeks you'll receive a copy of the article or book along with a bill for page charges for photocopying.

USING THE LITERATURE: TAKING NOTES

Earlier in this chapter we explained the system of using index cards for recording your references as you discovered them (see Figure 7.1). When you are ready to begin reading those sources, you use the same cards to take notes. Information from each source (e.g., journal article, encyclopedia entry) is recorded on the index card on which you wrote the author's name, title, etc. Use the back of the card if you need more space. Occasionally, you'll need even more room; make a continuation card, headed by the author's last name and a shortened title. If you often find yourself using two or more cards for each source, though, chances are you are not digesting the material very well. You are probably copying long quotations that you will never use. Save time by making the condensation immediately after you have read the piece while it is fresh in your understanding. Then write a summary of the major points.

Careful, responsible use of sources is the hallmark of the conscientious researcher. Inaccurate quotations and misleading summaries cause the reader to question the value of the entire piece of research. Whenever you use the words, ideas, data, or line of reasoning of another writer, you assume the responsibility of representing that author's work fairly and accurately and of crediting the source in the manner accepted by the discipline. See Chapter 8 for rules on documenting sources in the biological sciences.

The following passage, taken from an article entitled "Songs of Humpback Whales" (Payne and McVay 1971), represents the source material that a researcher might want to use for a monograph or term paper. Our subsequent treatment of summary, paraphrase, and quotation takes its examples from this passage.

> Humpback whales, like sperm whales, are found in all oceans of the world. However, while the sperm whale has been, and remains, the most numerous large cetacean on earth, the humpback has never been very plentiful. The principal concentration of humpback whales is in the Antarctic Ocean (5)[1], where they have probably never numbered more than 34,000 at any one time. However, the intense whaling of the past 40 years has reduced the number of humpbacks there to no more than a few percent of the original numbers.
>
> The International Whaling Commission has called for full protection of the humpback. Yet, even if this moratorium is honored, the number of humpbacks in the Southern Hemisphere seems dangerously low, perhaps too low to provide the pool of genetic variability needed to survive the next natural or manmade crisis.
>
> Though they have also been seriously overhunted in the Northern Hemisphere, small herds of humpbacks appear during natural periods of concentration (that is, for feeding, migration, delivering young, and the like). The waters near Bermuda are well known as such an area.

[1]Unlike most journals in the biological sciences, *Science*, which published the article from which this excerpt was taken, uses a numbering system of citation. Thus the (5) is keyed to a numerical list of sources at the end of the article. *Writer's Guide* employs the author-date system, explained in Chapter 8.

Summary

The most efficient method of note-taking, summary is economical. A good summary of an eleven-page article might be only a hundred words or so. Just how you summarize depends on the source and your intended use of it. In one case, the entire article might be relevant to your needs; then you would write a comprehensive summary that followed the structure of the original closely, even including some of the evidence. In another situation, you might be more concerned with getting just the central idea of the article; two or three sentences could capture that for you. Another time you might find only one section of the article relevant to your needs, so you summarize only that section. In all cases, use your own words when summarizing the work of the author. Only two rules apply to writing summaries: you must present the central idea of the original without distortion, and you must credit the source both in your text with a citation and at the end of the paper in your list of works cited.

Paraphrase

Paraphrase is the running restatement in your own words of the information or ideas of another writer. Unlike a summary, which greatly condenses the information, the rewording of a paraphrase may be nearly as long as the original. It also follows the same order of ideas as the original.

Original

> "The principal concentration of humpback whales is in the Antarctic Ocean (5), where they have probably never numbered more than 34,000 at any one time. However, the intense whaling of the past 40 years has reduced the number of humpbacks there to no more than a few percent of the original numbers." (Payne and McVay 1971)

Paraphrase

> Although the Antarctic Ocean contains the largest population of humpback whales, hunting pressure over the last 40 years has

reduced their numbers from a possible high of 34,000 to a small fraction of that (Payne and McVay 1971).

Quotation

Unlike paraphrase, quotation employs the exact words of the source. You must use great care to assure that the wording, spelling, punctuation, and data are precisely as they appear in the original. Although quotation is frequently used in the humanities and social sciences, the practice in biology is to avoid it whenever possible, partly because it is so uneconomical. A one-hundred-word quotation can be paraphrased in perhaps fifty to sixty words or summarized in twenty.

Quoted passages must be enclosed in quotation marks (“ ”). To delete material from a quotation, use ellipsis periods (. . .) to mark the deletion: “Humpback whales . . . are found in all oceans of the world.” (Payne and McVay 1971). To add material for purposes of clarification, use brackets ([ ]) to enclose the added words: “However, the intense whaling of the past 40 years has reduced the number of humpbacks there [the Antarctic Ocean] to no more than a few percent of the original numbers.” (Payne and McVay 1971).

As with summary and paraphrase, the writer must credit the source of a quotation. See Chapter 8 for details.

Photocopying

No matter how proficient you become at taking notes, you will find that you have to photocopy some things — a table, a figure, a particularly pertinent article. By all means take them to the photocopy machine and make a copy. Copyright laws permit such copying, provided you note the source on your copy. You will need this information when citing the material in your paper. Copy what you must, but use some common sense. Some students will plug rolls of coins into the copier rather than read anything in the library.

One Last Word

It is far better to read the material and take notes in the library than to check out all those books. Spending an hour or two in the library is more productive than carrying the books to your room. By the time you get there, you may be too tired to use them and they may sit for days.

USEFUL REFERENCE SOURCES IN THE LIFE SCIENCES

The following are only a few of the many resources available to the researcher in the biological sciences. Use this list of sources as an outline and add to it other items in your own library which you find useful.

Abstracting Services

Supply abstracts, allowing the researcher to do an initial screening for articles in the field of interest.

Environment Abstracts. EIC/Intelligence, New York. Contains a very helpful "How to use" section in the front of each issue. Reviews are classified by topic, e.g., Air Pollution, Food and Drugs. The Abstracts are accompanied by an Environment Index for ease of searching a topic.

Bibliographies and Guides

Bibliographies provide lists of books and articles in the field. Guides are how-to aids to research.

Kirk, Jr., T. G. 1978. *Library Research Guide to Biology: Illustrated Search Strategy and Sources.* Pierian Press, Ann Arbor, Mich. Helpful tips on everything from choosing the topic to using other guides. Appendix includes a library use quiz and numerous other helpful items.

Science for Society: a Bibliography. 1970–. American Association for the Advancement of Science Commission on Science Education, Washington, D.C.

Smith, R. C.; Reid, W. M. 1980. *Smith's Guide to the Literature of the Life Sciences.* 9th ed. Burgess, Minneapolis. Full of information on types of materials available in the library and how to use them.

Computer Search Services

The future in library searches. You ask for a specific type of reference, and the on-line computer searches and shows you everything it has listed on that topic. You may then request (depending on the capabilities of the service) a reference list, the abstracts of the articles you want, or complete copies of the articles.

Dialog, Information Services, California.

Bibliographic Retrieval Services (BRS)

Indexes

Used to identify subject headings and specific titles, some indexes provide reproductions of contents pages of journals, while others offer abstracts and related index sets.

Biological Abstracts. A collection of abstracts of articles from all over the world. Most comprehensive index. Generally at least one year behind the publications.

Biological Abstracts/RRM (formerly *Bioresearch Index*). An index to Biological Abstracts.

Biology Digest. A summary of selected articles from about 200 journals. Very readable summaries and articles. The abstracts are actually digests. "How to use" section in front of issue.

Current Contents. A reproduction of the table of contents of journals in selected fields. A quick way to scan a multitude of journals for contents.

Index Medicus. A major source of literature on medical topics.

Science Citation Index. Useful for tracing all references on a topic or by an author.

Dictionaries

Provide definitions, spelling, pronunciation, and usage of technical and scientific terms.

Abercrombie, M.; Hickman, C. J.; Johnson, M. L. 1973. *A Dictionary of Biology.* 6th ed. Penguin Books, Baltimore.

Gray, P. 1982. *The Dictionary of the Biological Sciences.* Krieger, New York.

McGraw-Hill Dictionary of the Life Sciences. 1976. McGraw-Hill, New York.

McGraw-Hill Dictionary of Scientific and Technical Terms. 2nd ed. 1978. McGraw-Hill, New York.

Encyclopedias

Useful sources of basic knowledge for the researcher new to the field. Do not end your research here and do not use as a cited source in a paper.

The McGraw-Hill Encyclopedia of Science and Technology. 4th ed. 1977. McGraw-Hill, New York. 15 volumes. Full of concise, useful, basic data covering the physical, earth, and life sciences, and engineering.

The McGraw-Hill Yearbook of Science and Technology. McGraw-Hill, New York. Annual supplement to update *The McGraw-Hill Encyclopedia of Science and Technology.*

Encyclopaedia Britannica. 1976. Encyclopaedia Britannica, Chicago. 30 volumes.

Britannica Yearbook of Science and the Future. 1969–. Encyclopaedia Britannica, Chicago. Annual update of the *Britannica.*

Handbooks

An indispensable source of methodology, frequently used tables of data, and taxonomic keys.

Altman, P. L.: Dittman, Dorothy S. 1966. *Environmental Biology.* Fed. Amer. Soc. Exp. Biol., Bethesda, MD. Excellent source for a quick reference check on environmental parameters.

Altman, P. L.; Dittman, Dorothy S. 1974. *Biology Data Book.* Fed. Amer. Soc. Exp. Biol., Bethesda, MD. 3 volumes.

Great source for information on such items as the gestation period of an animal, life expectancies.

American Public Health Association. 1985. *Standard Methods for the Examination of Water and Wastewater.* 16th ed. APHA, Washington, D.C. Describes all the accepted procedures for chemical and biological examination of water. Detailed descriptions of solution preparation, sample collecting and preservation.

Handbook of Chemistry and Physics. 1913–. Chemical Rubber, Cleveland, Ohio. Good reference book for chemical and physical data.

The Peterson Field Guide Series. Houghton Mifflin, Boston. A collection of identification booklets for use in the field or laboratory. Each treats one subject; e.g., fish, butterflies.

Informal Journals

Provide easy to read articles on many subjects.

Natural History. Carries a wide variety of articles ranging from art to science.

Science 86. The year changes annually. Contains short notes on current research areas as well as in-depth articles suitable for the general reader.

Smithsonian. Similar to *Natural History.* Many historical articles.

Review Journals

Reviews of current knowledge, often accompanied by a historical summary of research.

American Scientist. Good review articles in a variety of science areas.

Biological Reviews. In-depth review articles on biological topics.

Physiological Reviews. Technical review articles on physiological topics.

Quarterly Review of Biology. Excellent biology monographs.

Scientific American. Very readable articles on a wide variety of scientific topics.

Research Journals

Reports of current research in the field, written by professional researchers in technical language.

American Journal of Botany

American Zoologist

Biological Bulletin

BioScience

Developmental Biology

Ecology

Evolution

Immunology

Journal of Bacteriology

Journal of Biological Chemistry

Journal of Cell Biology

Journal of Experimental Biology

Journal of Morphology

Science

Yearbooks and Annuals

Provide yearly updating of research in the field.

Annual Review of Biochemistry, 1932–. Palo Alto, Calif.

Annual Review of Microbiology. 1947–. Palo Alto. Calif.

Cold Spring Harbor Symposia on Quantitative Biology. Cold Spring Harbor, L.I., New York. Each year's volume is devoted to an in-depth review of the research on one topic.

This is only a sampling of the resources available in your college's library. With practice and the assistance of helpful librarians you will become familiar with the holdings in the collection. Then, when you have found the material you need, Chapter 8 will guide you through the process of documenting your sources correctly.

[8] *Documentation of Sources*

PREVIEW: *Any time you use the words, data, or ideas of another researcher in your writing, you have to give credit. This chapter shows how to treat the most common situations.*

In-text citations
References cited
References to periodicals
References to books
Reference to abstracted source

Almost everyone embarking on a project of research and writing finds the issue of documentation confusing. Should you use footnotes or in-text citations? A **Bibliography** containing all the pertinent works, **References** noting all the works consulted in preparing the paper, or **References Cited** listing only those articles you cited in the paper? Which of the many formats should you follow? Turabian, MLA (Modern Language Association), CBE (Council of Biological Editors), APA (American Psychological Association) — all have their advocates. Because of its wide acceptance in biology, we have chosen to present the CBE format as detailed in the *CBE Style Manual* (1983). Before you begin a course project, though, you should consult your professor on choice of style. If you are writing for publication, check the journals in the field. Each editor spells out manuscript requirements and other guidelines for contributors; look near the table of contents. Pages 136–38 show the "Information for Contributors" page from *BioScience*.

Information for Contributors

Revised January 1986[1]

CORRESPONDENCE

Direct all correspondence to *BioScience* Editor, American Institute of Biological Sciences, 730 11th Street, NW, Washington, DC 20001-4584. Tel: 202/628-1500.

GENERAL INFORMATION

The editors welcome articles summarizing important areas of biological research, written for a broad audience of professional biologists and advanced students. *BioScience* also publishes articles on policy issues important to biologists, viewpoints, letters pertaining to material in *BioScience*, essays, and book reviews. In addition, The Biologist's Toolbox contains descriptions and reviews of instrumentation and computer hardware and software relevant to the professional biologist.

Papers are accepted for publication on the condition that they are submitted solely to *BioScience* and will not be reprinted or translated without the publisher's permission. All authors must transfer certain copyrights to the publisher.

The **Viewpoint** page and **Roundtable** essays may cover any topic of interest to biologists — from science policy and education to technical controversy. Viewpoints must not exceed two and a half double-spaced pages; essays must be no longer than ten double-spaced pages.

Articles must be no longer than **20–25 double-spaced pages,** including all figures, tables, and references. Scientific articles should review significant findings and include enough background for biologists in fields outside the author's; the writing should be as free of jargon as possible. The editors reserve the right to edit for style and clarity.

Submit an original and two copies of all manuscripts along with a cover letter. List in your cover letter the names of colleagues who have reviewed your paper plus the names, addresses, and telephone numbers of four potential referees from outside your institution but within North

[1]© American Institute of Biological Sciences. Reprinted with permission.

America. All articles are reviewed by outside referees and the editors for content and writing style.

Authors **must** obtain written permission to use in their articles any material copyrighted by another author or publisher. Include with your manuscript photocopies of letters granting permission; be sure credit to the source is complete.

About eight months usually elapse between initial receipt of a manuscript and publication.

MANUSCRIPT PREPARATION

Typing: Use **double-spacing throughout** all text, tables, references, and figure captions on one side only of 8½ × 11-inch white paper. Type all tables, figure captions, and footnotes on sheets separate from the text. Provide a separate title page with authors' names, affiliations, and addresses; include a sentence or two of relevant biographical information.

Style: Follow the *Council of Biology Editors Style Manual,* 5th ed., for conventions in biology except for references cited. For general style and spelling, consult *The Chicago Manual of Style,* 13th ed., and *Webster's Third International Dictionary.*

Abstract: Include an informative abstract no longer than 50 words; do not include a summary.

Measurement: All weights and measures must be in the metric system, SI units.

Symbols, acronyms: Define all symbols and spell out all acronyms the first time they are used.

Illustrations: Clear color transparencies will be considered for the cover. Photographs, maps, line drawings, and graphs must be camera-ready glossy black-and-white prints, photostats, or original art. On reverse, number and identify figures and indicate "top" of photographs. All **photographs must be untrimmed and unmounted,** 4 × 5 to 8 × 10 inches in size, and as clear as possible; photomicrographs should have a scale bar. Line drawings and graphs should be done by professional artists or scientific illustrators; those not done by professionals are rarely acceptable. Lettering must be large enough to be legible after a 50% reduction.

Footnotes: Keep footnotes to a minimum; number them with consecutive superscript numerals. Use symbols (see p. 79, *CBE Style Manual,* 5th ed.) for footnotes in tables. Reference citations referring to "personal communication" or unpublished data should be given as footnotes comprising source's name and affiliation and date.

References Cited: In-text citations must take the form (Author date); multiple citations should be listed in alphabetical order. Use first author's name and "et al." for in-text citations of works with more than two authors or editors; list every author or editor in the References Cited. All works cited in the text must be listed alphabetically in References Cited; works not cited in the text should not be listed. Follow the BIOSIS *List of Serials* for journal abbreviations, provide the full name of journals not listed there, and underline the titles of all books and journals. Refer to recent *BioSciences* for additional formatting; some examples:

A journal article: Bryant, P. J., and P. Simpson. 1984. Intrinsic and extrinsic control of growth in developing organs. *Quart. Rev. Biol.* 59:387–415.

A book: Ling, G. N. 1984. *In Search of the Physical Basis of Life.* Plenum Press, New York.

Chapter in book: Southwood, T. R. E. 1981. Bionomic strategies and population parameters. Pages 30–52 in R. M. May, ed. *Theoretical Ecology.* Sinauer Associates, Sunderland, MA.

Technical report: Lassiter, R. R., and J. L. Cooley. 1983. Prediction of ecological effects of toxic chemicals; overall strategy and theoretical basis for the ecosystem model. EPA-600/3-83-084. National Technical Information Service PB 83-261-685, Springfield, VA.

Unpublished paper: O'Leary, D. S. 1982. Risks and benefits of cooperating with the media. Paper presented at the annual meeting of the American Association for the Advancement of Science, Washington, DC, 8 January 1982.

In the following pages you will find explanations and examples of the most common types of citations and references. However, no brief treatment is likely to answer all of your questions about manuscript preparation and documentation. If you don't find an answer to your question here, consult the CBE manual, which is probably in your library's reference section. When all else fails ask your professor.

Generally the use of footnotes is discouraged. Include a note only if you must add essential information which would distract from the text if included there. Notes should be numbered consecutively throughout the paper. Type raised arabic numerals (like this[1]) to call attention to a note. Note what *BioScience* has to say about footnotes. Some journals, *Science* for one, use this system for both footnotes and citations. If you are told to use this system consult a copy of *Science* for the proper style. Consult with your professor about whether the footnotes are to go at the bottom of each page or on a separate page, headed **Footnotes.**

IN-TEXT CITATIONS

When using the words, data, or ideas of another writer, you credit the source by means of a citation. This enables the reader to locate the source in the list of references at the end of your paper. The CBE citation style is easy to learn and simple to use. The author's last name and the year of publication are inserted parenthetically in the text. Some journals (and your professor) may ask for the page numbers as well. It is useful to include page numbers if you have used a paraphrase or summary of a passage or a direct quotation. If there are two authors, list both last names. If there are more than two authors, list the name of the first author followed by **et al.** before the year.

Remember, citations are to sources you have actually read. If you must cite an article or book which was cited in a source you read, see one of following examples which will illustrate how to handle such a situation.

Placement of Citations

It seems that most students first learn documentation by putting all of their citations at the end of the paragraphs:

```
. . . have not been as well fed as the queen (Scott
1958).
```

Or at the end of sentences:

```
. . . of social behavior (Lindauer 1961). In such
a . . .
```

In both cases the manner of citation is correct — author and year in parentheses. Other ways to introduce the same information include

```
A 1958 article by Scott indicated . . .
Scott's study of bees (1958) showed . . .
A 1958 study of bees (Scott) shows . . .
```

A Work by Two Authors

Cite both names each time the reference occurs.

```
Olson and Wilder (1961) improved the quality . . .
. . . of the samples (Olson and Wilder 1961).
```

A Work by More Than Two Authors

Cite the surname of the first author followed by "et al."

```
This was confirmed by Brown et al. (1984).
Brown et al. (1984) confirmed that . . .
. . . was confirmed (Brown et al. 1984).
```

Two other acceptable formats for group authorships (but used less frequently) are "and co-workers" and "and others" after the first surname.

```
Brown and co-workers (1984) confirmed . . .
. . . was confirmed (Brown and others 1984).
```

A Work by an Institutional Author

Authorship of a work is sometimes attributed to a society, a government agency, or some other institution. In this case, the institution is cited as the author.

```
(CBE 1983)
(Council of Biological Editors 1983)
```

Common abbreviations may be used or an abbreviation established in the first citation of your paper, provided the meaning is clear.

(UNESCO 1980) [common abbreviation]
(American Public Health Association [APHA] 1985)
[first citation]
(APHA 1985) [subsequent citations]

A Work with No Author Given

When a book or article appears without an author's name, use either the title or "Anonymous."

```
("Instructions to Authors" 1982)
(Anonymous 1986a)
```

The "a" after 1986 refers to the fact that you have cited more than one "Anonymous" article from 1986. Use this same convention if you have to cite the same name for more than one source from the same year.

Two or More Works within Parentheses

Works by the same author(s) are arranged in order of date of publication.

```
(Oppenheimer 1960, 1963)
```

Works by different authors are arranged alphabetically by surnames.

```
(Benson and Parker 1961, Richards and Benson 1961).
```

Personal Communication (letters, interviews, phone conversations)

```
Senator Patrick Leahy (personal communication, Jan-
uary 12, 1986)
```

Personal communications are not included in the list of references because they cannot be consulted by the reader.

Government Documents

```
(USDA Forest Service 1976)
```

Statutes

```
(VT Environmental Laws and Regulations 1977)
     or
(VT Env Laws and Regs 1977)
```

From a Secondary Source

Sometimes a source is not available because it is out of print, or not in your library, or is in a foreign language which you do not read. It is still important to reference the original authors while at the same time being honest with your readers about just where you actually saw the material.

Read about it in another source

```
"Darwin (1859, cited by Leakey 1979) wrote . . ."
```

This citation indicates that you did not read Darwin's original publication but rather the 1979 book by Leakey. If possible include both sources in your References Cited section. Be sure to note, in parentheses after the Darwin citation, that you obtained the information from Leakey 1979. Use the same format for foreign language articles cited in another author's English article.

Abstract as a source

If your information about a particular reference is based on an abstract from *Biological Abstracts* or a similar source, cite the author(s) and year as explained above. In your Reference Cited section be sure to indicate your source was the abstract. See the following section on References Cited for a specific example.

REFERENCES CITED

The list of references at the end of an article or paper identifies the sources used, enabling the reader to locate the various articles and books. It is customary to call this list **References Cited** and to include every reference cited in the paper and no other. Your professor may, however, instruct you to use either a **Bibliography** or **References** list. A **Bibliography** lists not only the sources you used but as many additional pertinent articles as you can find. The **References** list, on the other hand, gives both the articles cited and additional sources you consulted but did not end up citing in the paper. As a general rule when you were preparing the paper, you compiled a Bibliography of many sources, selected from this list those References useful in preparing the paper, but found you needed to cite only a few of these articles to support your ideas or present your position. The last became your **Reference Cited** section. This distinction is becoming blurred and the terms **References, References Cited,** and **Literature Cited** are now used interchangeably.

Although each part of a paper should be done with great care, compiling the list of references calls for exacting precision. A misspelled name, an incorrect page number, an omitted date, a citation in the text missing from your list — any of these may trouble the reader and call into question the accuracy of the research and the researcher (you!).

The list of references begins on a separate page. The words **References Cited** are centered at the top. The entries are listed alphabetically by the author's last names or, in the case of institutional authorship, by the first significant word of the name. (If you had to use the numbering system each citation would be numbered in the order it appeared in the text and listed in sequence.) Use hanging indentation: the first line of each entry begins at the left margin, subsequent lines are indented two spaces. Capitalize only the first word of the title, the subtitle of books and articles, and proper nouns when they appear in scientific names. (Remember that in the binomial nomenclature only the Genus name is capitalized, not the species epithet.) Underline only those words which should be italicized, for example scientific names (`Felis domestica`). Do not underline titles of books, journals, or articles. For journals give the issue number only if each issue begins on page one.

If you made good notes on your reference cards when you first started to collect information for your paper, then preparing your References Cited section will be easy. First arrange in alphabetical (or numerical) order all the reference cards to be cited. Then type your entries.

The main parts of a reference to a journal article are **Author(s). Year. Title of the article. Abbreviated journal title. Volume (issue number if used). Inclusive pages.**

The main parts of a reference to a book are **Author(s). Year. Title and subtitle. Edition number (if not original). Place of publication. Name of publisher. Page numbers (if cited).**

If there is more than one author, list all with surname before the initials and separate each author by a semicolon. The year may be placed after the authorship or at the end of the citation. Whichever position you use, be sure it is the same for each entry. We prefer the year after the author. If the journal title is one word do not abbreviate; e.g., Science, BioScience. If you are uncertain about how to abbreviate a title refer to *BIOSIS list of serials: with CODEN, title abbreviations, new, changed and ceased titles. Philadelphia: BioSciences Information Service; (published each Janu-*

ary). Or, and this is our best method, refer to the journal itself to see how references citing its previous articles are abbreviated. (If you read this section before starting your research, keep this tip in mind when making your reference cards.)

The following are some of the more common types of reference entries you will be making. For entries which do not fit these examples refer to the *CBE Style Manual* or to a journal example.

References to Periodicals

Journal Article by One Author

Koch, A. L. 1971. The adaptive responses of Escherichia coli to a feast and famine existence. Adv. Microb. Physiol. 6: 147–217.

Journal Article by Two Authors

Broecker, W. S.; Peng, T. H. 1974. Gas exchange rate between sea and air. Tellus 26:21–35.

Journal Article by More Than Two Authors

Hansen, M.; Busch, L.: Burkhardt, J.; Lacy, W. B.; Lacy, Laura R. 1986. Plant breeding and biotechnology. BioScience 36(1):29–39.

Magazine Article

Wertenbaker, W. 1974. Profiles (Maurice Ewing – I). The New Yorker (Nov 4): 54–118.

Newspaper Article

Wasowicz, Lidia. 1984. Death of the Dinosaurs; Was extinction slow or sudden? The Chicago Tribune. Oct. 28; Sect 6:1–2.

Entire Issue of a Journal

Weaver, K. F., ed. 1981. Energy: A special report. Nat. Geogr. (Feb.):1–115.

Monograph

Tolbert, N. E. 1974. Photorespiration. Bot. Monogr. 10:474–504.

Periodical Published Annually

Cold Spring Harbor Symposia on Quantitative Biology. 1984. Recombination at the DNA level. Vol. 49. NY: Cold Spring Harbor Laboratory.

References to Books

Book by One Author

Desmond, A. J. 1977. The hot-blooded dinosaurs; a revolution in palaeontology. NY: Warner Books, Inc.

Book by Two or More Authors

Miller, J.; Van Loon, B. 1982. Darwin for beginners. NY: Pantheon Books.

Book by an Institutional Author

APHA. 1985. Standard methods for the examination of water and wastewater. 16th ed. Washington, DC: American Public Health Association.

Edited Book

Shepard, P.; McKinley, D., eds. 1969. The subversive science; essays toward an ecology of man. Boston: Houghton Mifflin Co.

Chapter in a Book

Brown, F. A. Jr. 1973. Biological rhythms. In: Prosser, C. L., ed. Comparative Animal Physiology. 3rd ed. Philadelphia: Saunders Co.

Reference to Abstracted Source

Calandra, S.: Quartaroli, G. C.; Montaguti, M. 1975. Effect of cholesterol feeding on cholesterol biosynthesis in maternal and foetal rat liver. Eur J Clin Invest 5(1):27–31. Taken from: Biol. Abst. 60:1394 (abstract no. 13047).

[9] A Concise Guide to Usage

PREVIEW: *In this chapter we present nine rules of effective writing, chosen because they are so often violated by the unwary writer. Read through the chapter to familiarize yourself with the material, then refer to it again as you revise and proofread your work.*

Usage is the name given to matters of correctness or suitability of language — as simple as that. Most of us learn standard usage from parents, friends, teachers, newspapers and books, radio and television. But we all have lapses and weak points that can be distracting to our readers. That is the reason for this section of the *Writer's Guide*. The rules explained below cover the most common questions of usage. Mastering them will not make you one of the world's great prose stylists, but it will help you to write more clearly and without the distractions that errors of usage can cause your reader.

RULE 1. SUBJECT AND VERB MUST AGREE IN NUMBER.

In English, nouns (and pronouns) and verbs are either singular or plural. If the subject noun (or pronoun) is singular, then the verb of the sentence must also be singular.

> Ellen (She) [**singular subject**] swims. [**singular verb**]
> The girls (They) [**plural subject**] swim. [**plural verb**]

So far, so good. But sentences like those aren't the ones that give writers problems. The difficulty surfaces when you write a sentence like this one:

> The value placed on a clean river by the two cities differ drastically.

Does that look all right to you? Let's see: the subject is *value,* singular; the verb is *differ,* plural. Subject and verb do not agree. This kind of error is common, especially in speaking, where it is easy to make. You hear the noun nearest to the verb, *cities,* and create a plural verb to match. Of course, *cities* is not the subject; *value* is. The correct sentence reads:

> The value placed on a clean river by the two cities differs drastically.

A simple test for complicated sentences is to omit everything but subject and verb, then look and listen:

> The value . . . differs. . . .

Collective nouns name a group or collection: *herd, club, nation, team,* etc. They take a singular verb if unity is stressed or a plural verb if their plurality is emphasized:

> The faculty *is* empowered to revise biology offerings.
>
> but
>
> The faculty *are* divided on the issue of required courses.

RULE 2. A PRONOUN MUST AGREE IN NUMBER WITH ITS ANTECEDENT.

If the antecedent (the noun that the pronoun replaces) is singular, the pronoun must be singular. If the antecedent is plural, the pronoun must be plural:

> Some students [**plural antecedent**] fail to submit their [**plural pronoun**] papers on time.

This would seem an easy rule to follow, yet mistakes are common. One student's explanation of how to teach windsurfing contained this sentence:

> Let the learner practice until they feel quite comfortable.

Here the subject (*learner*) is singular, but the pronoun (*they*) is plural. One way of correcting this sentence is to make the noun plural:

> Let learners practice until they feel quite comfortable.

This solution avoids the problem of gender introduced by the alternative:

> Let the learner practice until he (she? he or she?) feels quite comfortable.

RULE 3. USE THE CORRECT FORM OF THE PRONOUN.

The common personal pronouns (*I, me, he, him, she, her, it, we, us, you, they, them*) seldom cause much difficulty. Many writers do have problems with the punctuation of two classes of possessive pronouns. Never use an apostrophe with these forms:

possessive forms (act as modifiers)

my	*my* pen
your	*your* books
his	*his* belt
her	*her* car
its	*its* clarity
our	*our* house
your	*your* camera
their	*their* papers

substantive forms (act as nouns)

mine	That pen is *mine.*
yours	Which books are *yours?*
his	The brown belt is *his.*
hers	The second car is *hers.*
its	Of the wines tested for clarity, *its* is best.
ours	The yellow house is *ours.*
yours	What kind of camera is *yours?*
theirs	The papers on the desk are *theirs.*

Note: it's is a contraction of *it is.*

RULE 4. DON'T SHIFT VERB TENSES UNNECESSARILY.

Traditionally, writers in some fields use only the past tense of verbs, treating all events and ideas as if they occurred in the past. Writers in other fields may sometimes use the historical present, treating past events as if they were happening now:

> Shakespeare frequently alternates scenes of terror and tragedy with moments of comic relief.

Use whichever tense best suits your needs. Just be consistent: don't shift from past to present to past without a purpose.

RULE 5. PLACE MODIFIERS AS CLOSE AS POSSIBLE TO WORDS MODIFIED.

Writers-in-training are more apt to violate this rule with multi-word modifiers:

> Mangy and flea-bitten, I saw the dog sitting on my front steps.
>
> Our agency rents cars to salespeople of all sizes.
>
> Bouncing off parked cars, he spotted the driverless truck.

The meaning is clarified by placing the modifiers next to the words described:

> I saw the mangy and flea-bitten dog sitting on my front steps.
>
> Our agency rents cars of all sizes to salespeople.
>
> He spotted the driverless truck bouncing off parked cars.

RULE 6. WRITE COMPLETE SENTENCES.

A **sentence** is a group of words that contains a subject and a verb and expresses a complete thought. This is a sentence:

> My shoe is tight.

This is not:

> Because my shoe is tight.

Why not? What's the difference? Each group of words has a subject, *shoe,* and a verb, *is.* The only difference between the utterances is the addition of the word *because* to the second. The reason that "Because my shoe is tight" is not a sentence is that it doesn't express a complete thought; it cannot function as an independent unit. Read it aloud and you'll see what I mean. The listener (reader) is left dangling — because my shoe is tight *what?*[1]

Ironically, by adding a word, *because,* to the sentence, we've made it less than complete. This kind of word is called a **subordinator.** One kind of subordinator is the **relative pronoun:** *which, that, who, whom, what,* and *whose* are examples. The **subordinating conjunction** is a second kind. Common subordinating conjunctions are *because, after, when, although, as, before, if, unless, until, when, where.* The effect of adding these subordinators to a clause is to make that clause dependent:

"Because my shoe is tight" is an example of one kind of sentence fragment. It doesn't express a complete thought, it cannot stand alone. It must be attached to a complete sentence, like this:

> My foot hurts because my shoe is tight.

The sentence above has two **clauses.** Because the first clause, *my foot hurts,* expresses a complete thought and can stand alone, it is called **independent.** Because the second clause does not express a complete thought and cannot stand alone, it is called **dependent.**

A complete sentence, then, must contain an independent clause. It may contain additional elements as well.

complete sentence (independent clause):

> Stan stopped smoking recently.

complete sentence (two independent clauses and coordinating conjunction):

> Stan stopped smoking recently, and he feels healthier.

[1]Speech and writing have different requirements. In the following conversation, "because my shoe is tight" may function perfectly well: "Why are you limping?" "Because my shoe is tight."

complete sentence (dependent clause and independent clause):

Since Stan stopped smoking recently, he feels healthier.

RULE 7. AVOID COMMA SPLICE AND RUN-ON.

When independent clauses are joined, you must separate them with a comma plus *and, but, or, for, nor,* or *yet;* or with a colon; or with a semicolon. Violations of this rule are the comma splice and the run-on sentence.

wrong	The fluorescent light over the desk in my office isn't working, it hasn't worked since the painters were here. (comma splice)
correct	The fluorescent light over the desk in my office isn't working, and it hasn't worked since the painters were here.
correct	The fluorescent light over the desk in my office isn't working; it hasn't worked since the painters were here.

Note: See Chapter 10 for use of the colon.

RULE 8. DISTINGUISH BETWEEN HOMOPHONES.

Homophones are words pronounced alike but different in spelling and meaning. Using any of them incorrectly marks your writing as less than meticulous. You should master these common ones:

their, there, they're

their is a possessive pronoun:

on their own, their books

there has three common uses:

1. as an adverb meaning *in, at,* or *to that place:*

   She is going to build an addition there.

2. as a noun meaning *that place:*

   We live near there.

3. as a function word to introduce a clause:

   There are only two choices in the matter.

   they're is a contraction of *they are:*

   They're my best friends.

to, too, two

to is a preposition meaning *toward, as far as, until, etc.* With a verb, it is a sign of the infinitive:

The road to Jeffersonville is closed.

The second shift is from three to eleven.

The plant manager likes to play squash. (infinitive)

too is an adverb meaning *also, more than enough:*

The report was late too.

Too many cooks spoil the broth.

two is the number between one and three, used as an adjective or a pronoun:

"Two hamburgers, please."

Only two survived.

than, then

than is a conjunction used in comparisons:

She is taller than her brother.

then may be an adverb, adjective, or noun related to time:

I'm going to the meeting too. I'll see you then.

Since then he hasn't smiled.

RULE 9. AVOID SEXUALLY-BIASED LANGUAGE.

In recent years we have become much more aware of the ways language shapes our thinking. Most people realize that referring to Italians as "wops," for instance, not only demeans them but also makes it difficult for us to perceive Italians as anything but stereotypes.

I don't believe you're likely to practice racial or national stereotyping in your writing. But you and I along with millions of other Americans do practice another kind of linguistic bias nearly every time we write. I'm talking about sexual bias. Let me show you what I mean:

> Pioneers moved West, taking their wives and children with them.

What's wrong with that sentence? What's wrong is the assumption that the pioneers, the builders of our nation, were all males and that women (and children, for that matter) went along for the ride. That is simply not true. It is this kind of bias, perhaps unconscious, perhaps unintentional, that you need to watch for in your writing.

To be honest, avoiding sexist language isn't always easy. Because the English language lacks a singular pronoun that means *he or she,* the writer constantly has to deal with gender choices like these:

> When the shopper wishes to cash a check, she (he?). . . .
>
> Each student should write his (or her?) name at the top.

As a writer, you do have options:

1. Alternate female and male pronouns

   A sprinter warms up by stretching her muscles. A pianist runs over scales and chords to limber his fingers.

2. Rewrite to use the plural

   Sprinters warm up by stretching their muscles. Pianists run over scales and chords to limber their fingers.

3. Rewrite to avoid gender pronouns

   A sprinter warms up by stretching. A pianist runs over scales and chords.

A Final Word

It is impossible in these few pages to anticipate all the questions you might have as you write your papers. Every writer should have a copy of a handbook of usage. Buy one and refer to it as you revise your papers. A few minutes spent in this way can make all the difference on the impact of your paper.

[10] *Make Punctuation Work for You*

PREVIEW: *Correctly used, punctuation aids the reader's understanding of your writing. Incorrectly used, punctuation can confuse or misinform. This chapter focuses on the most common uses of each mark of punctuation. Read through the chapter to familiarize yourself with the material, then refer to it again as you revise and proofread your work.*

Punctuation marks clarify the meaning of our writing. Some usages are purely conventional: the colon (:) after "Dear Sir" in a business letter, for instance. Others have been established to express relationships between ideas. Your primary goal in punctuating should always be clarity of expression. Although common sense will often help you select the correct usage, there is no substitute for knowing a few basic rules.

Comma ,

The comma is the most frequently used, and abused, mark of punctuation. Relatively weak as a separator, it is less emphatic than the colon, semicolon, or dash. It indicates the briefest of pauses. Although there are dozens of uses of the comma, we'll look at only the most common.

To separate items in a series:

> The standard personal computer consists of memory, video display, keyboard, disc drive, and printer.

Note: a comma is used before the *and.*

To set off interrupters:

> The paleontologist's name for worms, *Vermes,* is really a catch-all term. (appositive)

The phylum Cnidaria, on the other hand, can also be called Coelenterata. (parenthetical expression)

Note: Interrupters are enclosed by a *pair* of commas.

To set off a long introductory phrase or clause:

In the deep snows at the top of the mountain, they hid a cache of supplies.

If you want to learn to ski the right way, you should take lessons.

To separate independent clauses joined by and, but, or, nor, for, yet:

Matt was interested in the job, but he didn't want to move away from his family.

The purchasing department ordered new furniture, and the office manager had the rooms painted.

To introduce a short quotation:

The librarian told them, "If you have a question, ask someone."

Semicolon ;

The semicolon provides more separation than the comma, less than the period. Its most common use is to separate independent clauses not joined by *and, but, or, nor, for, yet,* when you wish to show close relationship between those clauses. Otherwise, use a period.

The dean wanted a new curriculum; the faculty did not.
My mother was understanding the first time; she was upset the second time; the third time she was furious.

To Review: To show the degree of relationship between independent clauses, you have three options: semicolon, period, and comma with coordinating conjunction (*and, but, or, nor, for, yet*).

The birds have been arriving since mid-April; they have not built a single nest.

The birds have been arriving since mid-April. They have not built a single nest.

The birds have been arriving since mid-April, **yet** they have not built a single nest.

Colon :

The colon is used primarily to introduce a word, phrase, or clause that fulfills or explains an idea in the first part of the sentence. It is also used after the salutation of a business letter, to introduce a list, and to separate the title and subtitle of a book. Because it is a strong mark of punctuation, use only as directed.

To introduce or fulfill:

In that respect, Canada is like the United States**:** both have large numbers of non-English speakers.

On his deathbed the old miser made only one request**:** that his gold be buried with him.

To introduce a list:

The ethological study focussed on three factors**:**
age, sex, and dominance.

Note: Do not use a colon directly after a verb.

wrong	On her trip to France she visited: Paris, Chartres, and Mont St. Michel.
correct	On her trip to France she visited Paris, Chartres, and Mont St. Michel.

After the salutation of a business letter:

Dear Mrs. Irving**:**

To separate the title and the subtitle of a book:

*Biological Clocks***:** *Two Views*

Apostrophe ’

The apostrophe has three distinct uses: to mark the omission of one or more letters or numerals, to mark the possessive case, and to mark certain plurals.

To mark the omission of a letter or letters:

wouldn’t	(would not)
can’t	(cannot)
you’ll	(you will)
I’m	(I am)
it’s	(it is)
they’re	(they are)

To mark the omission of one or more numerals:

a ’57 Chevy	a 1957 Chevy
the summer of ’42	the summer of 1942

To form the possessive of a singular or plural noun not ending in s:

girl	girl’s
laboratory	laboratory’s
men	men’s
children	children’s

To form the possessive of a plural noun ending in s:

girls	girls’
books	books’
laboratories	laboratories’

To form the possessive of a singular noun of one syllable ending in s *or* s *sound:*

William James	William James’s philosophy
Brahms	Brahms’s First Symphony

To form the possessive of a singular noun of more than one syllable ending in s *or* s *sound:*

Socrates Socrates' school

Note: Do not use an apostrophe with possessive pronouns: *his, hers, yours, ours, theirs, whose, its* (*it's* means *it is*)

Parentheses ( )

Parentheses are used to enclose explanatory material within a sentence when such material is incidental to the main thought. Commas may also be used for this purpose; they are less formal and indicate a closer relationship to the main sentence than parentheses. Some writers use parentheses for the same purpose.

Senator Arndt (who just happens to be my brother-in-law) wrote the new farm credit bill.

Of his many novels (he wrote more than thirty), *Stairway to Darkness* was his favorite.

Note: Parentheses have special uses in citations. See the section on documentation.

Brackets [ ]

Brackets are marks of punctuation with limited but specific uses, especially in academic writing. Often when you excerpt part of a longer quotation, the meaning is not entirely clear. You may add clarification in brackets.

"The President [Truman] was determined that war policy be made by civilians, not generals."

"Freud's division of the psyche [id, ego, superego] has been disputed by many in recent years."

When you wish to acknowledge without changing an error in the quoted material, enclose the Latin word *sic* (thus) in brackets:

"The carriage careened wildly through muddy ruts until it broke an axel [sic]."

Note: Many typewriters do not have brackets. You can ink them in by hand. Do not use parentheses instead of brackets.

Ellipsis . . .

The omission (ellipsis) of part of a quoted passage is indicated by ellipsis marks: three spaced periods. Use these marks when you are quoting a long passage but wish to omit material.

> "These matriarchal tribes . . . often fight small wars to extend their territory."

When you delete the end of a sentence, use four periods:

> "Four score and seven years ago our fathers brought forth on this continent, a new nation. . . . Now we are engaged in a great civil war, testing whether that nation, or any nation so conceived and so dedicated, can long endure."

Dash —

The dash is probably the most overused mark of punctuation. Because it is so emphatic, its misuse stands out glaringly and is viewed as the sign of an overemotional style. Employ the dash only as described below.

To show an abrupt break in thought:

> "I explained all that to you yesterday when — oh, but that wasn't you."

To introduce a word or words for emphasis:

> You have only one choice — do it!

To separate a final summarizing clause from the preceding idea:

> Food, clothing, shelter, and fuel — these are all that Thoreau claimed are needed to sustain life.

Note: To type a dash, use two hyphens (--). Do not leave a space before or after the hyphens.

Quotation Marks " "

Quotation marks enclose the precise words spoken or written by someone other than the writer. Do not use them to identify indirect quotations or summaries.

To enclose direct quotations:

> In his book Bronsky asserts, "Mussolini's leadership was not entirely bad for Italy."
>
> Franklin Delano Roosevelt's powerful words, "The only thing we have to fear is fear itself," were spoken in the depths of national depression.

Note: When the quoted passage is embedded in a sentence, it is preceded by a comma (and followed by one if the sentence continues beyond the quotation).

Exception: Long quotations (more than four typed lines) are indented ten spaces and double spaced. Quotation marks are not used.

A quotation within a quotation uses single marks within double marks:

> In his inaugural speech Governor Harris urged her listeners to "Remember President Kennedy's message 'Ask not what your country can do for you — ask what you can do for your country.'"

Note: Periods and commas are placed inside the quotation marks. Semicolons and colons are placed outside. Question marks, exclamation points, and dashes are placed outside the quotation marks unless they are part of the original quotation.

To mark the titles of short stories, poems, essays, articles, and chapters of books, songs, symphonies, and plays in collections:

> Hemingway's story "The Snows of Kilimanjaro"
>
> Shelley's poem "To a Skylark"

Note: When they appear in the text, titles of books, full-length plays, magazines, and newspapers are *italicized* or *underlined,* not placed in quotation marks.

References

APA. 1983. Publication Manual of the American Psychological Association. 3rd ed. Washington, DC: American Psychological Association.

APHA. 1985. Standard methods for the examination of water and wastewater. 16th ed. Washington, DC: American Public Health Association.

Auth, J. T. 1971. Golgi Apparatus: Morphology and function. Senior thesis, Biology department, Saint Michael's College, Winooski, VT.

Barzun, J. 1971. A writer's discipline. In. On writing, editing, and publishing. Chicago. Univ. Chicago Press.

Brown, Jr. F. A.; Hastings, J. W.; Palmer, J. D. 1970. The biological clock: two views. New York: Academic Press.

CBE. 1983. Council of Biological Editors style manual. 5th ed. Bethesda, MD: CBE.

Desmond, A. J. 1977. The hot-blooded dinosaurs. New York: Warner Books.

Dombroski, Jean. 1979. Histological observations of dermal pad changes during growth hormone support of limb regeneration in the hypophysectomized newt, *Triturus (Notophthalmus, Diemictylus) viridescens*. Senior research thesis, Biology department, Saint Michael's College, Winooski, VT.

Gould, J. L. 1975. Honey bee recruitment; The dance-language controversy. Science 189:685–693.

Lorenz, K. 1963. On aggression. New York: Harcourt, Brace and World.

Lovelock, J. E. 1979. Gaia, a new look at life on earth. New York: Oxford Univ. Press.

Masse, Ann. 1980. Shelburne bay water quality study. A proposal to the National Science Foundation from the Biology department, Saint Michael's College, Winooski, VT.

Morgan, Elaine. 1972. The descent of woman. New York: Stein and Day.

Moroney, M. J. 1968. Facts from figures. Baltimore, MD: Penguin Books.

Morris, D. 1967. The naked ape. New York: McGraw-Hill.

Payne, R. S.; McVay, S. 1971. Songs of humpback whales. Science 173:585–597.

Swaminathan, M. S. 1984. Rice. Sci. Am. 250(1):80–93.

von Frisch, K. 1967. Dance language and orientation of bees. Cambridge, MA: Harvard Univ. Press.

Watson, J. D. 1968. The double helix. New York: New American Library.

Wenner, A. M. 1967. The bee language controversy. Boulder, CO: Educational Products Improvement Corp.

Young, Louise B. 1973. Power over people. New York: Oxford Univ. Press.

Index

1 2 3 4 5 6 7 8 9 0